SpringerBriefs in Applied Sciences and Technology

Computational Intelligence

Series Editor

Janusz Kacprzyk, Systems Research Institute, Polish Academy of Sciences, Warsaw, Poland

SpringerBriefs in Computational Intelligence are a series of slim high-quality publications encompassing the entire spectrum of Computational Intelligence. Featuring compact volumes of 50 to 125 pages (approximately 20,000–45,000 words), Briefs are shorter than a conventional book but longer than a journal article. Thus Briefs serve as timely, concise tools for students, researchers, and professionals.

More information about this subseries at http://www.springer.com/series/10618

Simão Moraes Sarmento · Nuno Horta

A Machine Learning based Pairs Trading Investment Strategy

Springer

Simão Moraes Sarmento
Instituto de Telecomunicações, IST
University of Lisbon
Lisbon, Portugal

Nuno Horta🆔
Instituto de Telecomunicações, IST
University of Lisbon
Lisbon, Portugal

ISSN 2191-530X   ISSN 2191-5318   (electronic)
SpringerBriefs in Applied Sciences and Technology
ISSN 2625-3704   ISSN 2625-3712   (electronic)
SpringerBriefs in Computational Intelligence
ISBN 978-3-030-47250-4   ISBN 978-3-030-47251-1   (eBook)
https://doi.org/10.1007/978-3-030-47251-1

This Springer imprint is published by the registered company Springer Nature Switzerland AG
The registered company address is: Gewerbestrasse 11, 6330 Cham, Switzerland

Contents

Acronyms

ADF	Augmented Dickey-Fuller
ANN	Artificial neural network
ARMA	Autoregressive–moving-average
DBSCAN	Density-based spatial clustering of applications with noise
ETF	Exchange-traded fund
FNN	Feedforward neural network
GARCH	Generalized autoregressive conditional heteroskedasticity
LSTM	Long short-term memory
MAE	Mean absolute error
MDD	Maximum drawdown
MLP	Multilayer perceptron
MSE	Mean squared error
OPTICS	Ordering points to identify the clustering structure
PCA	Principal component analysis
RNN	Recurrent neural network
ROI	Return on investment
SR	Sharpe ratio
SSD	Sum of Euclidean squared distance
t-SNE	T-distributed Stochastic Neighbor Embedding

Chapter 1
Introduction

1.1 Topic Overview

Pairs Trading is a well-known investment strategy developed in the 1980s. It has
been employed as one important long/short equity investment tool by hedge funds
and institutional investors Cavalcante et al. [2], and is a fundamental topic in this
work. This strategy comprises two steps. First, it requires the identification of two
securities, for example two stocks, for which the corresponding prices series display a
similar behaviour, or simply seem to be linked to each other. Ultimately, this indicates
that both securities are exposed to related risk factors and tend to react in an identical
way. Figure 1.1 illustrates how this behaviour can be found in some popular stocks.
In Figure 1.1a, we may observe how the price series from two car manufacturers
seem to be tied to each other. The same behaviour is also illustrated in Figure 1.1b,
this time illustrating the price series of two of the biggest retail stores in the United
States. Two securities that verify an equilibrium relation between their price series
can compose a pair.[1]

    Once the pairs have been identified, the investor may proceed with the strategy's
second step. The underlying premise is that if two securities' price series have been
moving close in the past, then this should persist in the future. Therefore, if an irreg-
ularity occurs, it should provide an interesting trade opportunity to profit from its
correction. To find such opportunities, the spread[2] between the two constituents of
the pairs must be continuously monitored. When a statistical anomaly is detected,
a market position is entered. The position is exited upon an eventual spread correction. It is interesting to observe that this strategy relies on the relative value of two
securities, regardless of their absolute value.

    We proceed to introduce how the strategy may be applied using an example from
this work. A more formal description concerning the trading setup is presented in

---

[1]Each constituent of a pair is sometimes referred to as a pair's leg.

[2]For now, the spread is defined as the difference between two securities' price series.

© The Author(s), under exclusive license to Springer Nature Switzerland AG 2021
S. Moraes Sarmento and N. Horta, *A Machine Learning based Pairs Trading
Investment Strategy*, SpringerBriefs in Computational Intelligence,
https://doi.org/10.1007/978-3-030-47251-1_1

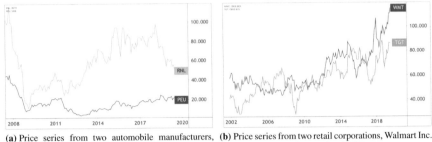

(a) Price series from two automobile manufacturers, Groupe Renault (RNL) and Peugeot S.A (UG).

(b) Price series from two retail corporations, Walmart Inc. (WML) and Target Corporation (TGT).

Fig. 1.1 Price series which could potentially form profitable pairs

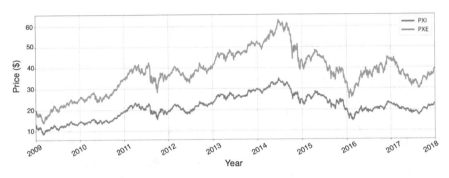

Fig. 1.2 Price series of two constituents of a pair during 2009–2018

Sect. 2.3. For now, we assume that two securities (identified by the tickers PXI and PXE) have been previously identified as forming a potential pair. PXE and PXI are two different securities that track indices related to energy exploration in the United States, and thus it is not surprising that their prices tend to move together. To confirm that this is the case, the two price series, from 2009 to 2018, can be observed in Fig. 1.2. The investor may calculate the mean value of the spread formed by the two constituents of the pair, as well as its standard deviation. These values describe the statistical behaviour known for that pair and which the investor expects to remain approximately constant in the future.

In the subsequent period, the spread, defined as $S_t = \text{PXI}_t - \text{PXE}_t$, is normalized[3] and cautiously monitored, as illustrated in Fig. 1.3. Although it evolves around its mean, it displays some noticeable deviations. Depending on their magnitude, they may trigger a trade. For that purpose, the investor defines the long and short thresholds, which define the minimum required deviation to open a long or short position, respectively. A long position presumes the spread will widen since its current value is below expected. Therefore, it entails buying PXI and selling PXE. Contrarily, a short

---

[3]Normalization, in this case, corresponds simply to subtract the mean and divide by the standard deviation.

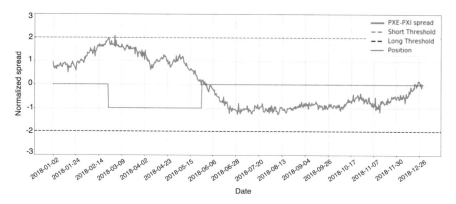

Fig. 1.3 Exemplifying a Pairs Trading strategy execution

position presumes the spread will narrow, and thus the opposite transaction takes place. The positions are liquidated when the spread reverts to its expected value.

Revisiting Fig. 1.3, we can identify the market position by the green line, which takes the values -1, 0 or 1 depending on whether the current position is short, outside the market, or long, respectively. In the second half of February 2018, we witness the opening of a short position, when the spread deviates across the short threshold, and its successive closure when the spread reverts back to zero.

The strategy just described presents diverse advantages. One incentive is overcoming the arduous process of characterizing the securities from a valuation point of view, which is a fundamental step in deciding to sell the overvalued securities and to buy the undervalued ones. By focusing on the idea of relative pricing, this issue is mitigated. Furthermore, this strategy is extremely robust to different market conditions. Regardless of the market direction (going up, down or sideways), if the asset the investor bought is performing relatively better than the one he sold, a profit can be made.

1.2 Objectives

Research in the field focuses mainly on traditional methods and statistical tools to improve the critical aspects of this strategy. Although in recent years Machine Learning techniques have gained momentum, reported Machine Learning based research for Pairs Trading in specific is sparse and lacks empirical analysis Krauss [4]. This leaves an opportunity to explore the integration of Machine Learning at different levels of a Pairs Trading framework.

The success of a Pairs Trading strategy highly depends on finding the right pairs. But with the increasing availability of data, more traders manage to spot interesting pairs and quickly profit from the correction of price discrepancies, leaving no margin

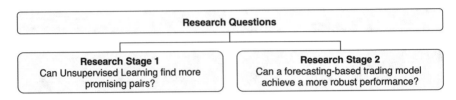

Fig. 1.4 Research questions to be answered

for the latecomers. If profitable pairs that are not being monitored by so many traders could be found, more lucrative trades would come along. The problem is that it can be extremely hard to find such opportunities. On the one hand, if the investor limits its search to securities within the same sector, as commonly executed, he is less likely to find pairs not yet being traded in large volumes. If on the other hand, the investor does not impose any limitation on the search space, he might have to explore an excessive number of combinations and is more likely to find spurious relations. To solve this issue, this work proposes the application of Unsupervised Learning to define the search space. It intends to group relevant securities (not necessarily from the same sector) in clusters, and detect rewarding pairs within them, that would otherwise be harder to identify, even for the experienced investor. We thus aim to answer the following: "Can Unsupervised Learning find more promising pairs?".

One of the disadvantages typically pointed out to a Pairs Trading strategy is its lack of robustness. Since the spread might still diverge after opening a market position, which counts on its reversion, the investor may see its portfolio value decline while the spread does not converge. In this work, a trading model which makes use of time series forecasting is proposed. It intends to define more precise entry points in order to reduce the number of days the portfolio value declines due to diverging positions. For this purpose, the potential of Deep Learning is evaluated. Therefore, the second objective of this work is to respond to the following question: "Can a forecasting-based trading model achieve a more robust performance?"

In the process of answering the research questions presented in Fig. 1.4, this work provides two additional contributions to the literature. First, it analyzes the suitability of ETF linked with commodities in a Pairs Trading setting. A total of 208 ETFs are considered. Secondly, it examines the application of a Pairs Trading strategy using 5-minutes frequency price series. This is particularly relevant given that most studies in the Pairs Trading literature use daily prices, with few exceptions Bowen et al. [1], Dunis et al. [3], Miao [5], Nath [6], Salvatierra and Patton [7]. Simulations are run during varying periods, between January 2009 and December 2018.

1.3   Outline

This book is composed by a total of seven chapters. Chapter 2 introduces the background and state-of-the-art concerning Pairs Trading along with some critical mathematical concepts to grasp the workings of this strategy. Chapters 3 and 4 describe the two implementations proposed in this work. The former describes in detail the proposed pairs selection framework (Research Stage 1), and the latter proposes a trading model that makes use of time series forecasting (Research Stage 2). Chapter 5 comprises some practical information concerning the way this investigation is designed. The results obtained are illustrated in Chap. 6. Finally, Chap. 7 focuses on answering the two research questions that motivated this research work, emphasizing the contributions made.

References

1. Bowen D, Hutchinson MC, O'Sullivan N (2010) High-frequency equity pairs trading: transaction costs, speed of execution, and patterns in returns. J Trading 5(3):31–38
2. Cavalcante RC, Brasileiro RC, Souza VL, Nobrega JP, Oliveira AL (2016) Computational intelligence and financial markets: a survey and future directions. Expert Syst Appl 55:194–211
3. Dunis CL, Giorgioni G, Laws J, Rudy J (2010) Statistical arbitrage and high-frequency data with an application to eurostoxx 50 equities. Liverpool Business School, Working paper
4. Krauss C (2017) Statistical arbitrage pairs trading strategies: review and outlook. J Econ Surv 31(2):513–545
5. Miao GJ (2014) High frequency and dynamic pairs trading based on statistical arbitrage using a two-stage correlation and cointegration approach. Int J Econ Financ 6(3):96–110
6. Nath P (2003) High frequency pairs trading with us treasury securities: risks and rewards for hedge funds. Available at SSRN 565441
7. Salvatierra IDL, Patton AJ (2015) Dynamic copula models and high frequency data. J Empir Financ 30:120–135

Chapter 2
Pairs Trading—Background and Related Work

2.1 Mean-Reversion and Stationarity

Before delving into more specific details of Pairs Trading, there exist a set of tools that should be introduced for the full comprehension of the related formalities. We start by introducing the concept of stationarity. A stochastic process is said to be stationary if its mean and variance are constant over time Adhikari and Agrawal [1]. That said, a stationary time series is mean reverting in nature and fluctuations around the mean should have similar amplitudes. A stationary process can also be referred to with respect to its order of integration $I(d)$, in which case a stationary time series is said to be an $I(0)$ process, whereas a non-stationary process is $I(1)$. The order of integration d is a summary statistic which reports the minimum number of differences required to obtain a stationary series.

Stationary series are particularly interesting for financial traders who take advantage of the stationary property by placing orders when the price of a security deviates considerably from its historical mean, considering the price will revert back. However, they are rarely found in the set of financial time series. In fact, price series are often non-stationary, and thus, in theory, completely unpredictable. This pattern should remind the reader of the spread illustrated in Fig. 1.3. This is not a coincidence. As we will see, Pairs Trading is particularly interesting because it provides a way of obtaining an artificial stationary time series from the combination of two non-stationary time series.

It should be clarified that mean-reversion and stationarity are not the same mathematical concept. According to Chan [5], the mathematical description of a mean-reverting time series is that the change of the time series in the next instant is proportional to the difference between the mean and the current value. Whereas, the mathematical description of a stationary time series is that the variance of the logarithmic value of the time series increases slower than that of a geometric random

© The Author(s), under exclusive license to Springer Nature Switzerland AG 2021
S. Moraes Sarmento and N. Horta, *A Machine Learning based Pairs Trading Investment Strategy*, SpringerBriefs in Computational Intelligence,
https://doi.org/10.1007/978-3-030-47251-1_2

walk.[1] That is, its variance is a sub-linear function of time, rather than a linear function, as in the case of a geometric random walk. Some necessary tools to detect and model the aforementioned properties are now presented in detail.

2.1.1 Augmented Dickey-Fuller Test

The ADF test is a hypothesis test designed to assess the null hypothesis that unit root is present in a time series, proposed by Dickey and Fuller [7]. The test assumes stationarity as the alternative hypothesis. A consecutive change in time series can be modelled as

$$\Delta y(t) = \lambda y(t-1) + \mu + \beta t + \alpha_1 \Delta y(t-1) + \cdots + \alpha_k \Delta y(t-k) + \epsilon_t, \quad (2.1)$$

where $\Delta y(t) \equiv y(t) - y(t-1)$, μ is a constant, β is the coefficient on a time trend and k is the lag order of the autoregressive process. The ADF test evaluates whether $\lambda = 0$. If the hypothesis is rejected, and $\lambda \neq 0$, the change of a time series at time t depends on the value of the series at time $t-1$, meaning that the series cannot be a simple random walk. The test statistic associated is the regression coefficient λ (calculated with $y(t-1)$ as the independent variable and $\Delta y(t)$ as the dependent variable) divided by the standard error (SE) of the regression fit, $\frac{\lambda}{SE}$. Note that when mean reversion is expected, $\frac{\lambda}{SE}$ has a negative value. This value should be compared with the critical values corresponding to the distribution of the test statistic, which can be found tabulated and used to decide whether the hypothesis can be accepted or rejected at a given probability level.

2.1.2 Hurst Exponent

The Hurst exponent can be used to evaluate the stationarity of a time series. Thus, according to the definition presented in Sect. 2.1, this metric assesses whether the time series' speed of diffusion from its initial value is slower than that of a geometric random walk Hurst [19]. Formally, the speed of diffusion can be characterized by the variance as

$$\text{Var}(\tau) = \langle |z(t+\tau) - z(t)|^2 \rangle, \quad (2.2)$$

where $z(t)$ is the logarithmic time series value at time t, τ is an arbitrary time lag and $\langle \cdot \rangle$ is the average across time. Having defined an indicator for the speed of diffusion,

---

[1]The geometric random walk is the default stochastic process model used to describe stock market data. Describing this model falls out of the scope of this work, but the interested readers may refer to Vempala [25].

it is still necessary to define the geometric random walk standard speed, by which to compare. We may assume that its speed of diffusion can be approximated as

$$\langle |z(t + \tau) - z(t)|^2 \rangle \sim \tau^{2H}. \tag{2.3}$$

The symbol H represents the Hurst exponent. For a price series exhibiting a geometric random walk, $H = 0.5$, and (2.3) simply becomes

$$\langle |z(t + \tau) - z(t)|^2 \rangle \sim \tau. \tag{2.4}$$

Yet, as H decreases towards zero, the speed of diffusion reduces, meaning the price series is more mean-reverting. Contrarily, as H increases toward 1, the price series is increasingly trending. The Hurst exponent can thus be interpreted as an indicator of the degree of mean-reversion or trendiness.

2.1.3 Half-Life of Mean-Reversion

The half-life of mean-reversion is a measure of how long it takes for a time series to mean-revert. To calculate this metric, the discrete-time series from (2.1) can be transformed into its differential form, so that the changes in prices now become infinitesimal quantities. Additionally, the term β may be ignored since we are dealing with price series and the constant drift in price, if any, tends to be of a much smaller magnitude than its daily fluctuations. If, for further simplification, the lagged differences are also ignored in (2.1), the expression can be written as

$$dy(t) = (\lambda y(t - 1) + \mu)dt + d\varepsilon. \tag{2.5}$$

This expression describes an Ornstein-Uhlenbeck process, where $d\varepsilon$ represents some Gaussian Noise Chan [5]. Assuming λ has been calculated as described in Sect. 2.1.1, the analytical solution for the expected value of $y(t)$ is represented by

$$E(y(t)) = y_0 \exp(\lambda t) - \mu/\lambda(1 - \exp(\lambda t)). \tag{2.6}$$

From the analysis of (2.6), we may confirm that for $\lambda > 0$ the time series will not be mean reverting, as stated in Sect. 2.1.1. For a negative value of λ, the expected value of the time series decays exponentially to the value $-\frac{\mu}{\lambda}$ with the half-life of decay equals to $\frac{log(2)}{\lambda}$. This result implies that the expected duration of mean reversion is inversely proportional to the absolute value of λ, meaning that a larger absolute value of λ is associated with a faster mean-reversion. Therefore, $\frac{log(2)}{\lambda}$ can be used as a natural time scale to define many parameters on a trading strategy and avoiding brute-force optimization of a free parameter.

2.1.4   Cointegration

Cointegration was first proposed in an article, by two econometricians, Engle and Granger [14]. A set of variables is said to be cointegrated if there exists a linear combination of those variables, of order d, which results in a lower order of integration, $I(d-1)$. In the context of this work, cointegration is verified if a set of $I(1)$ variables (non-stationary) can be used to model an $I(0)$ variable (stationary). Formally, considering two time series, y_t and x_t which are both $I(1)$, cointegration implies there exist coefficients, μ and β such that

$$y_t - \beta x_t = u_t + \mu, \tag{2.7}$$

where u_t is a stationary series. This is particularly interesting as it provides a way to artificially create a stationary time series that can be used for trading. Finding a time series with the same properties amid raw financial time series is an extremely hard task.

Tests for cointegration identify stable, long-run relationships between sets of variables. The most common cointegration tests are the Engle-Granger two-step method and the Johansen test. The two-step Engle-Granger cointegration test proceeds as follows:

1. Test if a unit root is present in the series y_t and x_t using ADF test. If affirmative, proceed to step 2.
2. Run regression defined in (2.7), using Ordinary Least Squares, and save the residuals, $\hat{u}_t$.
3. Test the residuals $\hat{u}_t$ for a unit root, using ADF test (or similar).
4. If the null hypothesis of a unit root in the residuals (null of no-cointegration) is rejected, meaning the residual series is stationary, the two variables are cointegrated.

A major issue with the Engle-Granger method is that the choice of the dependent variable may lead to different conclusions, as pointed by Armstrong [2]. The Johansen test is particularly useful for detecting cointegrating vectors when working with more than two securities, but that falls out of the scope of this work.

Having described the fundamental mathematical concepts required to understand the workings of a Pairs Trading strategy, we proceed to disclose how they can be applied in practice. As described in Sect. 1.1, a Pairs Trading strategy is usually organized in two stages, the identification of the pairs and the trading itself. Following this line of reasoning, the two next sections will go over each stage separately.

2.2 Pairs Selection

The pairs selection stage encompasses two steps: (i) finding the appropriate candidate pairs and (ii) selecting the most promising ones.

Starting with (i), the investor should select the securities of interest (e.g. stocks, futures, ETFs, etc.) and from there start searching for possible pair combinations. In the literature, two methodologies are typically suggested for this stage: performing an exhaustive search for all possible combinations of the selected securities, or arranging them in groups, usually by sector, and constrain the combinations to pairs formed by securities within the same group. While the former may find more interesting unusual pairs, the latter reduces the likelihood of finding spurious relations. For example, Caldeira and Moura [3], Krauss et al. [21] impose no restriction on the universe from which to select the pairs. Contrarily, Do and Faff [8], Dunis et al. [11], Gatev et al. [15] arrange the securities on category groups and select pairs within the same group. Figure 2.1 illustrates these two techniques, in a two-dimensional reduced setting, with the characters indicating hypothetical categories.

With respect to (ii), the investor must define what criteria should be used to select a pair combination. The most common approaches to select pairs are the distance, the cointegration and the correlation approaches, which we proceed to describe in more detail.

2.2.1 The Minimum Distance Approach

The work from Gatev et al. [15] can be considered as the baseline for distance-based selection criteria. It proposes selecting pairs which minimize a distance criterion. The authors construct a cumulative total return index for each stock and normalize it to the first day of a 12-month formation period. This period is used for calculating

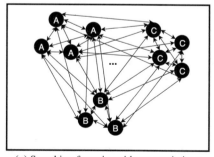 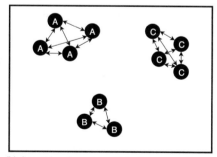

(a) Searching for pairs without restrictions. (b) Searching for pairs within the same category.

Fig. 2.1 Pairs search common techniques

the Sum of Euclidean Squared Distances (SSD) between the constructed time series. Then, the authors suggest ranking the pairs according to the minimum historic SSD.

However, according to Krauss [20], the choice of Euclidean squared distance as a selection metric is analytically sub optimal. To prove this statement, let $p_{i,t}$ and $p_{j,t}$ be realizations of the normalized price processes $P_i = (P_{i,t})_{t \in T}$ and $P_j = (P_{j,t})_{t \in T}$ of securities i and j composing a pair. Additionally, $s^2_{P_i - P_j}$ denotes the empirical spread variance and it is expressed as

$$s^2_{P_i - P_j} = \frac{1}{T} \sum_{t=1}^{T} \left(p_{i,t} - p_{j,t} \right)^2 - \left(\frac{1}{T} \sum_{t=1}^{T} \left(p_{i,t} - p_{j,t} \right) \right)^2 . \tag{2.8}$$

The average sum of squared distances, $\overline{ssd}_{P_i, P_j}$ in the formation period is then given by

$$\overline{ssd}_{P_i, P_j} = \frac{1}{T} \sum_{t=1}^{T} \left(p_{i,t} - p_{j,t} \right)^2 = s^2_{P_i - P_j} + \left(\frac{1}{T} \sum_{t=1}^{T} \left(p_{i,t} - p_{j,t} \right) \right)^2 . \tag{2.9}$$

An optimal pair according to the SSD criterion would be one that minimizes (2.9). But this implies that a zero spread pair across the formation period would be considered optimal. Logically, this is not consistent with the idea of a potentially profitable pair, since a good candidate should have high spread variance and strong mean-reversion properties to provide trading opportunities. As this metric does not account for this requirement, it is likely to form pairs with low spread variance and limited profit potential.

2.2.2 The Correlation Approach

The application of correlation on return levels to define a pairs' selection metric is examined in Chen et al. [6]. The authors intend to construct a robust pair selection metric that finds promising pairs without penalizing individual price divergences, as it happens in Gatev et al. [15]. Therefore, they use the same dataset as in Gatev et al. [15] for a direct comparison. The sample variance of the spread returns of a pair composed by asset i and j is defined as return on buy minus return on sell, as

$$s^2_{R_i - R_j} = s^2_{R_i} + s^2_{R_j} - 2 \hat{\rho}_{R_i, R_j} \sqrt{s^2_{R_i}} \sqrt{s^2_{R_j}}, \tag{2.10}$$

where $\hat{\rho}$ is the correlation found. It can be inferred from (2.10) that, although imposing high return correlation between two stocks' returns leads to lower variance of spread returns, the return time series of the individual stocks may still exhibit high variances.

The results using Pearson correlation show better performance, with a reported monthly average of 1.70% raw returns, almost twice as high as the obtained using the distance metric suggested by Gatev et al. [15]. However, it should be noted that, although correlation performed better than SSD as a selection metric, it is still not optimal, as two return level correlated securities might not share an equilibrium relationship, and divergence reversions cannot be explained theoretically.

2.2.3 The Cointegration Approach

The cointegration approach entails selecting pairs for which the two constituents are cointegrated. If two securities, Y_t and X_t are found to be cointegrated, then by definition the series resulting from the linear combination

$$S_t = Y_t - \beta X_t, \qquad (2.11)$$

where β is the cointegration factor, must be stationary. Naturally, defining the spread series this way is particularly convenient, since under these conditions it is expected to be mean-reverting. According to Krauss [20], cointegration approach for pairs selection is a more rigorous framework when compared to the distance approach. This is mainly because cointegration finds econometrically more sound equilibrium relationships. The selection criteria follows the procedure illustrated in Fig. 2.2.

The most cited author in this field is Vidyamurthy [26]. In his book, a set of heuristics for cointegration strategies are proposed. Huck and Afawubo [17] performs a comparison study between the cointegration approach and the distance approach using S&P 500 constituents, under varying parameterizations, considering risk loadings and transactions costs. The authors found that the cointegration approach significantly outperforms the distance method, which corroborates the hypothesis that the cointegration approach identifies more robust relationships.

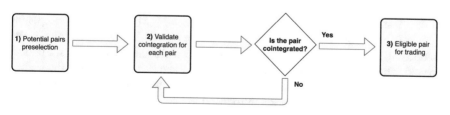

Fig. 2.2 Cointegration approach for selecting pairs

2.2.4 Other Approaches

Although the three approaches just described are the most commonly mentioned in the literature, there are naturally other methodologies used for tackling this issue. We find it relevant to mention the strategy introduced by Ramos-Requena et al. [23]. The authors suggest ranking the pairs based on the spread's Hurst exponent. They obtain superior results when compared with the classical approaches based on cointegration and correlation, especially as the number of pairs in the portfolio grows. This indicates this may be an interesting research direction.

The approaches described up to this point are statistically motivated. Nevertheless, this does not have to be the case. Some research work also focuses on specific pairs which tend to display an equilibrium that is theoretically founded. For instance, Dunis et al. [9] focus on the gasoline crack spread, which is defined as the differential between the price of crude oil and petroleum products extracted from it. In this scenario, there is an evident justification for the underlying equilibrium. The same happens with the corn/ethanol crush spread, which is examined in Dunis et al. [12]. Here, the link can be explained by the fact that corn is the principal raw ingredient used to produce ethanol.

2.3 Trading Execution

Having explored the most relevant techniques for selecting pairs, we proceed to examine the existent trading strategies. Unlike pairs selection, which can be found in diverse configurations, the trading strategy usually follows the threshold-based trading model described next.

2.3.1 Threshold-Based Trading Model

In the classical framework, proposed by Gatev et al. [15], the criteria for opening a trade is based on the spread divergence. If the spread between two price series composing a pair, diverges by more than two historical standard deviations, a trade is opened. The trade is closed upon convergence to the mean, at the end of the trading period or when delisting occurs. This model should be familiar to the reader as it corresponds to the implementation introduced in Sect. 1.1.

This model can be described more formally as follows:

i Calculate the spread's ($S_t = Y_t - X_t$) mean, μ_s and standard deviation, σ_s during the pair's formation period.
ii Define the model thresholds: the threshold that triggers a long position, α_L, the threshold that triggers a short position, α_S, and the exit threshold α_{exit} that defines the level at which a position should be exited.

iii Monitor the evolution of the spread, S_t and control if any threshold is crossed.
iv In case α_L is crossed, go long the spread by buying Y and selling X. If α_S is triggered, short the spread by selling Y and buying X. Exit position when α_{exit} is triggered and a position was being held.

It is important to note that the way the spread, S_t, is defined is tied to the technique used to identify the pairs. For example, when applying the minimum distance approach, the spread composed by the securities Y and X is simply defined as $S_t = Y_t - X_t$. However, when using cointegration, the spread is defined in a slightly different manner, as $S_t = Y_t - \beta X_t$. This subtlety affects the way positions are set. If the spread is defined as $S_t = Y_t - X_t$, then entering a long position entails buying Y and selling X in the same number of shares, or investing the same capital on both, in case the investor wants to assume a dollar-neutral position. However, if the spread is defined as $S_t = Y_t - \beta X_t$, one must decide how to handle the cointegration factor. Different alternatives have been proposed in the literature, and there seems to be no consensus on the most appropriate one. According to Chan [5], the investor should invest in one share of Y and β shares of X. Rad et al. [22] suggests the investor should invest \$1 in the long leg and determine the value in the short leg according to the cointegration factor. Other authors, such as Dunis et al. [11], simply neglect the cointegration factor to enforce a money-neutral position and invest the same amount in each leg.

2.3.2 Other Trading Models in the Literature

The strategy just described is not concerned with optimizing the market entry points. There is no guarantee that the point immediately after crossing the threshold is an optimal entry point. In fact, the spread might continue to diverge before converging again.

Some efforts have emerged trying to propose more robust models. Techniques from different fields, such as stochastic control theory, statistical process modelling and Machine Learning have been studied. Some examples in the literature include Elliott et al. [13], who model the spread with a mean-reverting Gaussian Markov chain, observed in Gaussian noise. The trading positions are set according to the relative difference between the series prediction and the observation. Dunis et al. [9] proposes a similar framework but this time based on Artificial Neural Networks. The results obtained by Machine Learning approaches have proved very promising, and are explored in the next section in more detail.

2.4 The Application of Machine Learning in Pairs Trading

Over the last years, more research has become available on the application of Machine Learning in the field of finance Cavalcante et al. [4]. With the advent of more computational power, it became easier to train complex Deep Neural Networks capable of generating very promising results. Nevertheless, the application of Machine Learning in Pairs Trading in specific is still scarce.

Dunis et al. [9, 10, 12] are the main authors applying Machine Learning to Pairs Trading, even though their work is limited to specific spreads. Dunis et al. [9] apply Artificial Neural Networks to model the gasoline crack spread. The authors train the neural networks to predict the percentage change in the spread, from one closing price to the next. The trading rules are then set by comparing the change forecast with fixed thresholds, according to,

$$
\begin{cases}
\text{if } \Delta \tilde{S}_t > X, & \text{open or stay long the spread} \\
\text{if } \Delta \tilde{S}_t < -X, & \text{open or stay short the spread}, \\
\text{if } -X < \Delta \tilde{S}_t < X, & \text{remain out of the spread}
\end{cases}
\tag{2.12}
$$

where $\Delta \tilde{S}_t$ is the model's predicted spread return and X is the value of the threshold filter (which the authors optimize in-sample). A Recurrent Neural Network (RNN) achieved profits of 15% on average before transaction costs. However, the impact of transaction costs is significant. Dunis et al. [12] apply Artificial Neural Networks in an adjusted framework to the corn/ethanol crush spread. In this case, the authors found more satisfying results with a profit after transaction costs of approximately 20%.

Huck [16] also explores the integration of Machine Learning in Pairs Trading, by introducing a combined approach, built on Artificial Neural Networks. The methodology uses RNNs to generate a one-week ahead forecast, from which the predicted returns are calculated. A ranking algorithm is then applied to extract the undervalued and overvalued stocks. In the trading step, the top stocks of the ranking are bought and the bottom stocks sold. The results obtained are, once again, very satisfactory.

Table 2.1 summarizes what we consider to be the most relevant work applying Machine Learning in Pairs Trading. In addition to the studies just described, we include some additional work in the field Huck [18], Thomaidis et al. [24].

Apart from the studies presented, Machine Learning techniques applied in Pairs Trading remain fairly unexplored. Furthermore, the existing limitations suggest there might be open research opportunities in this direction. Lastly, all the applications focus on the supervising power of Machine Learning and seem to neglect how Unsupervised Learning could also be advantageous in this field.

Table 2.1 Literature exploring the application of Machine Learning in Pairs Trading

Work	Data sample	Contribution	Limitation
Krauss et al. [21]	U.S S&P 500 1992–2015	Analyzes the effectiveness of deep neural networks, gradient-boosted-trees, and random forests in the context of statistical arbitrage	Equilibrium pairs are not identified. The authors buy the stocks most likely to outperform the market and sell the less likely to do so
Dunis et al. [9]	Gasoline crack spread 1995–2005	Explore the potential of using Deep Learning models to forecast the spread returns. The authors apply Multilayer Perceptron, High-order Neural Networks and Recurrent Neural Networks	The evidence is limited to specific spreads
Dunis et al. [10]	Corn/eth. crush spread 2005–2010		
Dunis et al. [12]	Soybean/oil crush spread 1995–2005		
Huck [16]	U.S S&P 100 1992–2006	Introduces a combination of forecasting techniques (Neural Networks) and multi-criteria decision making (MCDM) methods (Electre III) for ranking pairs	It is unclear if the satisfactory results are due to the combined forecasts or the MCDM
Huck [18]	U.S. S&P 100 1993–2006	Similar to [16]. In addition, it indicates that better results can be obtained when using multiple-step forecasts	
Thomaidis et al. [24]	Selected stocks 2005	Introduces a combination of neural network theory and financial statistics for the detection of statistical arbitrage opportunities	Further experimentation is needed to improve results when transaction costs are taken into account

2.5   Conclusion

In this section, we introduced the main research topics related to Pairs Trading. In Sect. 2.2, we started by reviewing the standard procedures to search for pairs, which include searching for pairs within the same sector or searching for all possible combinations. Next, we analyzed the selection methods, with an emphasis on the correlation, distance and cointegration approaches. In Sect. 2.3, we examined the threshold-based trading model and briefly motivated some more innovative trading techniques. Finally, we focused on the literature available, exploring the application of Machine Learning in Pairs Trading.

References

1. Adhikari R, Agrawal RK (2013) An introductory study on time series modeling and forecasting. arXiv:13026613
2. Armstrong JS (2001) Principles of forecasting: a handbook for researchers and practitioners, vol 30. Springer Science & Business Media, Berlin
3. Caldeira J, Moura GV (2013) Selection of a portfolio of pairs based on cointegration: a statistical arbitrage strategy. Available at SSRN 2196391
4. Cavalcante RC, Brasileiro RC, Souza VL, Nobrega JP, Oliveira AL (2016) Computational intelligence and financial markets: a survey and future directions. Expert Syst Appl 55:194–211
5. Chan E (2013) Algorithmic trading: winning strategies and their rationale, vol 625. Wiley, New York
6. Chen H, Chen S, Chen Z, Li F (2017) Empirical investigation of an equity pairs trading strategy. Management Science
7. Dickey DA, Fuller WA (1979) Distribution of the estimators for autoregressive time series with a unit root. J Am Stat Assoc 74(366a):427–431
8. Do B, Faff R (2010) Does simple pairs trading still work? Financ Anal J 66(4):83–95. https://doi.org/10.2469/faj.v66.n4.1
9. Dunis CL, Laws J, Evans B (2006) Modelling and trading the gasoline crack spread: a non-linear story. Deriv Use Trading Regul 12(1–2):126–145
10. Dunis CL, Laws J, Evans B (2009) Modelling and trading the soybean-oil crush spread with recurrent and higher order networks: a comparative analysis. In: Artificial higher order neural networks for economics and business, IGI Global, pp 348–366
11. Dunis CL, Giorgioni G, Laws J, Rudy J (2010) Statistical arbitrage and high-frequency data with an application to eurostoxx 50 equities. Liverpool Business School, Working paper
12. Dunis CL, Laws J, Middleton PW, Karathanasopoulos A (2015) Trading and hedging the corn/ethanol crush spread using time-varying leverage and nonlinear models. Eur J Financ 21(4):352–375
13. Elliott RJ, Der Van, Hoek* J, Malcolm WP, (2005) Pairs trading. Quant Financ 5(3):271–276
14. Engle RF, Granger CW (1987) Co-integration and error correction: representation, estimation, and testing. Econ: J Econo Soc 251–276
15. Gatev E, Goetzmann WN, Rouwenhorst KG (2006) Pairs trading: performance of a relative-value arbitrage rule. Rev Financ Stud 19(3):797–827
16. Huck N (2009) Pairs selection and outranking: An application to the s&p 100 index. Eur J Oper Res 196(2):819–825
17. Huck N, Afawubo K (2015) Pairs trading and selection methods: is cointegration superior? Appl Econ 47(6):599–613. https://doi.org/10.1080/00036846.2014.975417

18. Huck N (2010) Pairs trading and outranking: The multi-step-ahead forecasting case. Eur J Oper Res 207(3):1702–1716
19. Hurst HE (1951) Long-term storage capacity of reservoirs. Trans Am Soc Civil Eng 116:770–799
20. Krauss C (2017) Statistical arbitrage pairs trading strategies: review and outlook. J Econ Surv 31(2):513–545
21. Krauss C, Do XA, Huck N (2017) Deep neural networks, gradient-boosted trees, random forests: statistical arbitrage on the s&p 500. Eur J Oper Res 259(2):689–702
22. Rad H, Low RKY, Faff R (2016) The profitability of pairs trading strategies: distance, cointegration and copula methods. Quant Financ 16(10):1541–1558. https://doi.org/10.1080/14697688.2016.1164337
23. Ramos-Requena J, Trinidad-Segovia J, Sánchez-Granero M (2017) Introducing hurst exponent in pair trading. Phys A: Stat Mech Appl 488:39–45
24. Thomaidis NS, Kondakis N, Dounias G (2006) An intelligent statistical arbitrage trading system. In: SETN
25. Vempala S (2005) Geometric random walks: a survey
26. Vidyamurthy G (2004) Pairs trading: quantitative methods and analysis, vol 217. Wiley

Chapter 3
Proposed Pairs Selection Framework

3.1 Problem Statement

As the popularity of Pairs Trading grows, it is increasingly harder to find rewarding pairs. The scarcity of such promising pairs forces a search expansion to broader groups of securities, on the expectation that by considering a larger group, the likelihood of finding a good pair will increase. The simplest procedure commonly applied is to generate all possible candidate pairs by considering the combination from every security to every other security in the dataset, as it was represented in Fig. 3.1a. This results in a total of $\frac{n \times (n-1)}{2}$ possible combinations, where n represents the number of available securities.

Two different problems arise immediately. First, the computational cost of testing mean-reversion for all the possible combinations increases drastically as more securities are considered. In a setting where securities are constantly being monitored to form new pairs, this is especially relevant. The second emerging problem is frequent when performing multiple hypothesis tests at once, and is referred to as the multiple comparisons problem. By definition, if 100 hypothesis tests are performed (with a confidence level of 5%) the results should contain a false positive rate of 5%. To illustrate how this might be relevant, one might consider the hypothetical situation where 20 shares are being handled. There would be $\frac{20 \times 19}{2} = 190$ possible combinations, which means approximately 9 results would be wrong (5% of 190). Since finding real cointegrated pairs is uncommon among a randomly selected group of securities, this number is considerably high. Formally, the probability ($\bar{\alpha}$) of committing at least one type I error[1] when performing m independent comparisons is given by

$$\bar{\alpha} = 1 - (1 - \alpha)^m, \tag{3.1}$$

---

[1] In statistical hypothesis testing a type I error is the rejection of a true null hypothesis (false positive), while a type II error is the non-rejection of a false null hypothesis (false negative).

© The Author(s), under exclusive license to Springer Nature Switzerland AG 2021
S. Moraes Sarmento and N. Horta, *A Machine Learning based Pairs Trading Investment Strategy*, SpringerBriefs in Computational Intelligence, https://doi.org/10.1007/978-3-030-47251-1_3

where α represents the probability level of a single test. This is known as the family-wise error rate (FMER).

The multiple comparison problem cannot be eliminated entirely, but its impact can be mitigated. Two options consist of applying multiple correction tests like Bonferroni or running fewer statistical tests. Bonferroni is one of the most commonly used approaches for multiple comparisons. However, the Bonferroni correction is substantially conservative. In fact, Harlacher [8] found that Bonferroni correction turns out to be too conservative for pairs selection and impedes the discovery of even truly cointegrated combinations. The author recommends the effective pre-partitioning of the considered asset universe with the purpose of reducing the number of feasible combinations and, therefore, the number of statistical tests.

This aspect might lead the investor to pursue the usual, more restrictive approach of comparing securities only within the same sector, as previously illustrated in Fig. 3.1. This dramatically reduces the number of necessary statistical tests, consequently reducing the likelihood of finding spurious relations. Moreover, there are fundamental reasons to believe that securities within the same sector are more likely to move together as they are exposed to similar underlying factors. This method also has the advantage of being particularly easy to execute. However, the simplicity of this process might also turn out to be a disadvantage. The more traders are aware of the pairs, the harder it is to find pairs not yet being traded in large volumes, leaving a smaller margin for profit.

This disequilibrium motivates the search for a methodology that lays in between these two scenes: an effective pre-partitioning of the universe of assets that does not limit the combination of pairs to relatively obvious solutions, while not enforcing excessive search combinations.

3.2 Proposed Framework

In this work, we propose the application of an Unsupervised Learning algorithm, on the expectation that it infers meaningful clusters of assets from which to select the pairs. The motivation is to let the data explicitly manifest itself, rather than manu-ally defining the groups each security should belong to. The proposed methodology encompasses the following steps:

1. Dimensionality reduction—find a compact representation for each security;
2. Unsupervised Learning—apply an appropriate clustering algorithm;
3. Select pairs—define a set of rules to select pairs for trading.

The three following sections will explore in detail how each of the previous steps is conducted.

3.3 Dimensionality Reduction

In the quest for profitable pairs, we wish to find securities with the same systematic risk-exposure. According to the Arbitrage Pricing Theory,[2] these securities generate the same long-run expected return. Any deviations from the theoretical expected return can thus be seen as mispricing and serve as a guideline to place trades. To extract the common underlying risk factors for each security, the application of PCA on the return series is proposed, as described in Jolliffe [9].

PCA is a statistical procedure that uses an orthogonal transformation to convert a set of observations of possibly correlated variables into a set of linearly uncorrelated variables, the principal components. The transformation is defined such that the first principal component accounts for as much of the variability in the data as possible. Each succeeding component, in turn, has the highest variance possible under the constraint that it must be orthogonal to the preceding components. It is especially interesting to note that each component can be seen as representing a risk factor Avellaneda and Lee [3].

The PCA application goes as follows. First, the return series for a security i at time t, $R_{i,t}$, is obtained from the security price series P_i,

$$R_{i,t} = \frac{P_{i,t} - P_{i,t-1}}{P_{i,t-1}}. \tag{3.2}$$

Next, the return series must be normalized since PCA is sensitive to the relative scaling of the original variables. This is performed by subtracting the mean, $\overline{R}_i$, and dividing by the standard deviation σ_i, as

$$Y_i = \frac{R_i - \overline{R}_i}{\sigma_i}. \tag{3.3}$$

From the normalized return series of all assets, the correlation matrix ρ is calculated, where each entry is determined by

$$\rho_{ij} = \frac{1}{T - 1} \sum_{t=T}^{M} Y_{i,t} Y_{j,t}. \tag{3.4}$$

The role of matrix ρ will soon be made more clear. Note that the motivation for applying PCA on return series lies in the fact that a return correlation matrix is more informative for evaluating price co-movements. Using the price series instead could result in the detection of spurious correlations as a result of underlying time trends.

The next step consists of extracting the eigenvectors and eigenvalues, to construct the principal components. The eigenvectors determine the directions of maxi-

---

[2] Arbitrage pricing theory (APT) is a general theory of asset pricing that holds that the expected returns of a financial asset can be modelled as a linear function of various factors or theoretical market indices. For more information, we redirect the reader to Ross [12].

mum variance, whereas the eigenvalues quantify the variance along the corresponding direction. Eigenvalues and eigenvectors can be determined using singular value decomposition (SVD). For that purpose, we stack the normalized return series for all n securities in a matrix A. By direct application of the SVD theorem on matrix A, it is decomposed as

$$A = U S V^T. \tag{3.5}$$

Matrix U is orthogonal, and its columns are the left singular vectors. Matrix S is diagonal and contains the singular values arranged in descending order along the diagonal, from the eigenvalue corresponding to the highest variance to the least. Matrix V is the transposed orthogonal matrix, and its rows are the right singular vectors.

Note that by multiplying $A^T A$ we obtain $V S^2 V^T$. But $A^T A$ is exactly the correlation matrix ρ previously calculated, hence, V can be determined. In this way, it is possible to find the eigenvectors of the data matrix A. At this point, we select the k eigenvectors corresponding to the k directions of maximum variance, where k represents the number of features to describe the transformed data. The more eigenvectors are considered, the better the data is described. The matrix containing the selected eigenvalues in order of significance is called the feature vector.

Finally, the new dataset is obtained by multiplying the original matrix A by the feature vector, resulting in a matrix with size $n \times k$. Please note that the position (i, j) of the final matrix contains the result from the product between the normalized return series, i by the eigenvector j. The matrix obtained is now reduced to the selected k features.

To conclude this section, the number of features, k, needs to be defined. A usual procedure consists in analyzing the proportion of the total variance explained by each principal component, and then use the number of components that explain a fixed percentage, as in Avellaneda and Lee [3]. In this work, however, a different approach is adopted. Because an Unsupervised Learning algorithm is applied using these data features, there should be a consideration for the data dimensionality. High data dimensionality presents a dual problem. The first being that in the presence of more attributes, the likelihood of finding irrelevant features increases. The second is the problem of the curse of dimensionality. This term is introduced by Bellman [4] to describe the problem caused by the exponential increase in volume associated with adding extra dimensions to Euclidean space. This has a tremendous impact when measuring the distance between apparently similar data points that suddenly become all very distant from each other. Consequently, the clustering procedure becomes very ineffective. According to Berkhin [5], the effect starts to be severe for dimensions greater than 15. Taking this into consideration, the number of PCA dimensions is upper bounded at this value and is chosen empirically. The implications of varying this parameter can be seen in Sect. 6.2.3.1.

3.4 Unsupervised Learning

Having transformed each time series in a smaller set of features with the application of PCA, an Unsupervised Learning technique can be effectively applied. We proceed to describe the decision process followed in selecting the most appropriate algorithm.

3.4.1 Problem Requisites

There exist several Unsupervised Learning clustering techniques capable of solving the task at hand. Thus, some problem-specific requisites are first defined to constrain the set of possibilities:

- No need to specify the number of clusters in advance;
- No need to group all securities;
- Strict assignment that accounts for outliers;
- No assumptions regarding the clusters' shape.

In more detail, first, no assumptions should be taken concerning the appropriate number of clusters in the data given that there is no prior information in this regard. By making the number of clusters data-driven, we introduce as little bias as possible. Secondly, it should not be imposed that every security is assigned to a group because we should expect to find securities with very distinct price series and ultimately divergent PCA decompositions. From a clustering point of view, these securities will be identified as outliers. Therefore, we should ensure they do not interfere with the clustering procedure by selecting a clustering algorithm robust in dealing with outliers. In third place, the assignment should be strict otherwise the number of possible pair combinations increases, which is conflicting with the initial goal. Finally, because there is no prior information indicating the clusters should be regularly shaped, the selected algorithm should not adopt this assumption.

3.4.2 Clustering Methodologies

Having described the problem requisites, we may proceed to select an appropriate clustering type. A common scheme to refer to the different clustering algorithms consists of categorizing them in three broad groups: partitioning clustering, hierarchical clustering, and density-based clustering. The goal of this work is not to perform an exhaustive exploration of all the existent options but rather to find a sound approach consistent with the properties stated so far. Hence, we are not concerned with explaining all the different methodologies in detail but rather to provide some intuition on the decisions taken.

Partitioning algorithms, such as k-means MacQueen et al. [10], are not considered suitable mainly for three reasons. The first is that these algorithms do not handle well noisy data and outliers. Secondly, clusters are forced to have convex shapes. The assumption that the data follows a normal distribution around the centre of the cluster might be too strong of an approximation at this stage. Especially because the data-dimensionality precludes the visualization of its distribution in space. Lastly, the obligation of specifying the number of clusters in advance is also undesirable in a scenario under which the model should have as few parameters as possible. One could argue that by performing a grid search on the number of clusters, an optimal partition should correspond to the configuration that minimizes a given criterion. This way, specifying a range of clusters would be sufficient. However, it is not evident what criteria leads to the selection of the most promising pairs. Clustering integrity metrics, such as the silhouette coefficient Rousseeuw [13], sound appropriate. However, we empirically verified that greater cluster integrity is not necessarily strictly correlated with more promising pairs.

Hierarchical clustering algorithms have the advantage of providing a dynamic termination criteria. This lets the investor select a group displacement with the desired level of granularity. Yet, this might add unnecessary bias from the investor. As consequence, the selection can approximate the results obtained using standard search methods, which we aim to prevent. For this reason, hierarchical clustering would only be appropriate under the existence of an automatic termination criterion.

Finally, density-based clustering algorithms display some benefits in the context of this work. First, clusters can have arbitrary shapes, and thus no gaussianity assumptions need to be adopted regarding the shape of the data. Secondly, it is naturally robust to outliers because it does not group every point in the dataset. Lastly, it requires no specification of the number of clusters. Therefore we proceed to investigate how density-based clustering could be applied in this work.

3.4.3 DBSCAN

A density-based clustering algorithm interprets the clustering space as an open set in the Euclidean space that can be divided into a set of its connected components. The algorithm DBSCAN (Density-Based Spatial Clustering of Applications with Noise) Ester et al. [7], is the most influential in this category. According to the author, the reason why we can easily detect clusters of points is that there is a typical density of points within each cluster which is considerably higher than outside of the cluster.

To formalize this notion of clusters' dependence on density, the key idea relies on the concepts of density and connectivity. There are essentially two model parameters. They are ε and $minPts$. To understand the algorithm, we start by defining the underlying fundamental concepts, w.r.t ε and $minPts$.

Definition 3.1 ε-neighborhood: The ε-neighborhood of a point q is defined as

$$N_\varepsilon(q) = \{p \in X | d(q, p) \leq \varepsilon\}, \tag{3.6}$$

where $d(q, p)$ represents the distance between q and p, and X is the set of all points.

Definition 3.2 *Core point*: A point q is core point if it verifies

$$|N_\varepsilon(q)| \geq minPts,$$

where $|N_\varepsilon(q)|$ represents the number of points within the ε-neighborhood of q. Note also that $minPts$ accounts for the point itself.

Definition 3.3 *Directly density-reachable*: Point p is directly density-reachable from point q in a set of points X provided that $p \in N_\varepsilon(q)$ and q is a core point.

Definition 3.4 *Density-reachable*: A point p is density-reachable from a point q if there is a chain of objects $p_1, ..., p_n$, where $p_1 = q$ and $p_n = p$ such that p_{i+1} is directly density-reachable from p_i.

Definition 3.5 *Density-connected*: A point p is density-connected to a point q if both points are density-reachable from a common core point.

Figure 3.1 illustrates the concepts just defined, when the parameter $minPts$ takes the value of 5. Figure 3.1a describes a core point. Point q is a core point, given that is contains $minPts$ within the circle of radius ε, its ε-neighborhood. Point p_1 belongs to the ε-neighborhood of q. Like p_1, a point that is not a core point but is included in an ε-neighborhood is called a border point. A point that does not belong to any neighborhood, such as p_2, is considered an outlier.

Figure 3.1b illustrates the concept of directly density-reachable and density reachable points. Point p_1 is directly density-reachable from q. Furthermore, p_2 is density-reachable from q, as both q and p_1 are core points. However, q is not density reachable from p_2, because p_2 is not a core point. As just demonstrated, density-reachability is not symmetric in general. Only core objects can be mutually density-reachable.

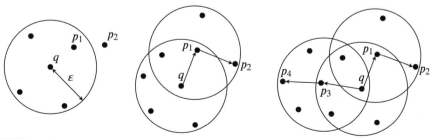

(a) Core points and outliers. **(b)** Density-reachable points.   **(c)** Density-connected points.

Fig. 3.1 DBSCAN illustration of basic concepts, with $minPts = 5$

Finally, Fig. 3.1c illustrates the concept of density connection. Points p_2 and p_4 are density-connected, since both points are density reachable from q. Note that density-connectivity is a symmetric relation.

We may now characterize a cluster. If p is a core point, then it forms a cluster together with all points (core or non-core) that can be reached from itself. This implies that each cluster contains at least one core point. Non-core points can also be part of a cluster, but they form its edge since they cannot be used to reach more points. Thus, a cluster satisfies two properties: (i) All points within a cluster are mutually density-connected and (ii) if a point is density-reachable from any point of the cluster, it must belong to the cluster. If a point does not belong to any cluster, it is considered noise. The DBSCAN execution can be described in a simplified manner as follows:

1. Find the points in the ε-neighborhood of every point and identify the core points with more than $minPts$ neighbours.
2. Find the connected components of core points on the neighbour graph, ignoring all non-core points.
3. Assign each non-core point to a nearby cluster if the cluster is an ε-neighbor, otherwise assign it to noise.

In spite of the DBSCAN advantages stated so far, there exist some caveats that should be noted. First, the algorithm is very sensitive to the parameters, namely the definition of ε. Moreover, even after finding a suitable ε, the algorithm assumes clusters are evenly dense. This results in an inability to find clusters of varying density. Although the parameter ε may be well adapted for one given cluster density, it might be unrealistic for another cluster with a different density, within the same dataset. This situation is depicted in Fig. 3.2. It is evident that cluster A, B and C could be found using the same ε. But in doing so, we would only identify cluster A and consequently neglect clusters A_1 and A_2. On the other hand, if we adapt ε to find A_1 and A_2, clusters B and C will be indistinguishable from noise.

The last drawback pointed out is particularly relevant in this work. Because there is a natural unbalance in the categories of the securities considered (described in Table 5.2), and assuming the category plays a relevant role in describing a security, then clusters are expected to have different densities. The expectation of unbalanced density clusters is a catalyst for moving on with OPTICS.

3.4.4 OPTICS

OPTICS, proposed by [1], addresses the problem of detecting meaningful clusters in data of varying density. The algorithm is inspired on DBSCAN, with the addition of two new concepts that we proceed to define. On the following definitions, we maintain the nomenclature used so far.

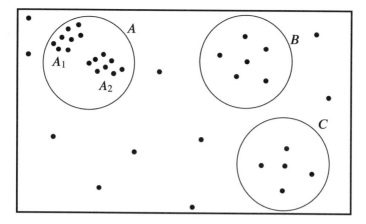

Fig. 3.2 Clusters with varying density

Definition 3.6 *Core-distance*: Let $minPts$-$distance(p)$ be the distance from a point p to its $minPts'$ neighbor. Then, the core-distance of p is defined as

$$core\text{-}dist_{\varepsilon,minPts}(p) = \begin{cases} \text{Undefined}, & \text{if } |N_\varepsilon(q)| < minPts \\ minPts\text{-}distance(p), & \text{otherwise} \end{cases} . \quad (3.7)$$

The core-distance of a point p is, simply put, the smallest distance ε' between p and a point in its ε-*neighborhood* such that p is a core point with respect to ε'. If p is not a core point in the first place, the core-distance is not defined.

Definition 3.7 *Reachability-distance*: The reachability-distance of p with respect to o, described as $reach\text{-}dist_{\varepsilon,minPts}(p, o)$ is defined as

$$reach\text{-}dist_{\varepsilon,minPts}(p, o) = \begin{cases} \text{Undefined}, & \text{if } |N_\varepsilon(o)| < minPts \\ max(core\text{-}distance(o), distance(o, p)), & \text{otherwise} \end{cases} .$$

$$(3.8)$$

The reachability-distance of a point p with respect to a point o can be interpreted as the smallest distance such that p is directly density-reachable from o. This requires that o is a core point, meaning that the reachability-distance cannot be smaller than the core-distance. If that was the case, o would not be a core point.

Figure 3.3 represents the concepts of core-distance and reachability-distance just described. The scenario represented defines $minPts = 5$ and $\varepsilon = 6\,mm$. On the left image, the core-distance, represented as ε', portrays the minimum distance that makes p a core point. For a radius of $\varepsilon' < \varepsilon$, there already exist five points ($minPts$) within the circle. The concept of reachability-distance is represented on the right image. The reachability-distance between q_1 and p must assure p is a core point, although the distance between the two points is actually smaller. The reachability-distance

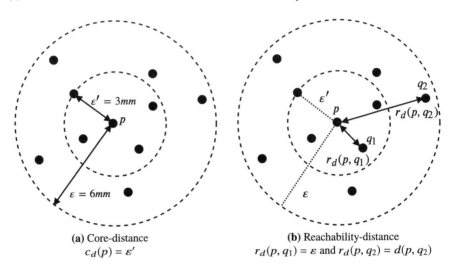

(a) Core-distance
$$c_d(p) = \varepsilon'$$

(b) Reachability-distance
$$r_d(p, q_1) = \varepsilon \text{ and } r_d(p, q_2) = d(p, q_2)$$

Fig. 3.3 Illustration of core-distance and reachability-distance

Fig. 3.4 Extracting clusters
using OPTICS

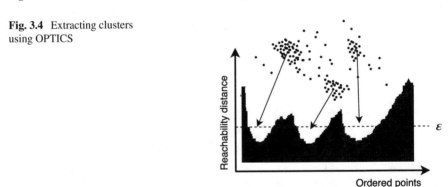

between q_2 and p, corresponds to the actual distance between the two points because any distance larger than ε' already assures p is a core point.

The OPTICS implementation is fundamentally equivalent to DBSCAN, with the exception that instead of maintaining a set of known, but so far unprocessed cluster members, a priority queue is used. We do not include the pseudo-code here, but the interested readers may refer to Ankerst et al. [1].

The algorithm returns the points and the corresponding reachability-distances. The points returned are ordered in such a way that spatially closest points become neighbours in the ordering. From this information, a reachability plot can be constructed, by arranging the ordered points in the x-axis and the reachability-distance on the y-axis, as shown in Fig. 3.4. Since points belonging to a cluster are closer from each other, they have a low reachability-distance to their nearest neighbour. As a consequence, the clusters are pictured as valleys in the reachability plot, which will be deeper for denser clusters.

Having described the reachability plot framework for cluster analysis, we may now describe the clustering extraction procedure. It can be performed mainly in two ways. The first is to inspect the reachability plot and fix a suitable ε on the y-axis, as exemplified in Fig. 3.4. This way, the chosen value is imposed for every cluster in the dataset, which simply yields the same results as applying DBSCAN with the same value for ε and $minPts$. The second alternative consists in letting an automatic procedure select the most suitable ε for each cluster. This is performed by defining the minimum steepness on the reachability plot that should constitute a cluster boundary. Ankerst et al. [1] describes in detail how this process can be accomplished, with the introduction of some additional new concepts. From this work's point of view, the key idea to retain is that this procedure addresses the problem of finding the most suitable ε for each cluster. As a byproduct, it makes the clustering process closer to being parameterless, as the investor only needs to specify $minPts$. Note that, in practice, the automatic cluster extraction procedure requires information regarding the minimum steepness on the reachability plot necessary to detect a cluster boundary. However, the default value proposed by the library implementation is adopted as it is independent of the data. Hence, the investor does not need to worry about this value.

3.5 Pairs Selection Criteria

Having generated the clusters of assets, it is still necessary to define a set of conditions for selecting the pairs to trade. It is critical that the pairs' equilibrium persists. To enforce this, we propose the unification of methods applied in separate research work. According to the proposed criteria, a pair is selected if it complies with the four conditions described next:

1. The pair's constituents are cointegrated.
2. The pair's spread Hurst exponent reveals a mean-reverting character.
3. The pair's spread diverges and converges within convenient periods.
4. The pair's spread reverts to the mean with enough frequency.

We proceed to go over each step with more detail to describe the reasoning behind. First, as described in Sect. 2.2.3, cointegration methods proved to achieve good performance in pairs selection, as cointegration based approaches identify econometrically more sound equilibrium relationships when compared to distance/correlation based methods. Thus, a pair is only deemed eligible for trading if the two securities that form the pair are cointegrated. To test this condition, we propose the application of the Engle-Granger test, due to its simplicity. One critic Armstrong [2] points out to the Engle-Granger test has to do with the fact that the choice of the dependent variable may lead to different conclusions. To mitigate this issue, we propose that the Engle-Granger test is run for the two possible selections of the dependent variable and that the combination that generated the lowest t-statistic is selected.

Secondly, an additional validation step is suggested to provide more confidence in the mean-reversion character of the pairs' spread. It aims to constrain false positives, possibly arising as an effect of the multiple comparisons problem. The condition imposed is that the Hurst exponent associated with the spread of a given pair is enforced to be smaller than 0.5, assuring the process leans towards mean-reversion. This additional step ensures that undesirable data samples that made it through the cointegration test but are not mean-reverting can now be detected and discarded. This criterion is adapted from the work of Ramos-Requena et al. [11], in which the authors claim that cointegration per se is too restrictive of a condition to impose and that is rarely fulfilled (something that is also mentioned in Chan [6]). For that reason, the authors propose ranking the pairs based on the Hurst exponent and choose them accordingly. The perspective assumed in this work is different, as we still impose that two legs forming a pair are cointegrated, but we use the Hurst exponent value as a second check to look for mean-reverting properties.

It should be noted that the Hurst exponent measures the degree of mean-reversion of a time series, and not its stationarity. Strictly speaking, this means that it cannot be interpreted as a direct replacement of the cointegration test. Although in practice, most stationary time series are also mean-reverting, there may be exceptions. As an example, we urge the reader to consider the process X which verifies $X_t = X_{t-1}$ for $t > 0$. Let X_0 take the value 1 with probability 0.5 and 0 otherwise. X is stationary but not mean reverting, proving stationarity does not imply mean-reversion. Nevertheless, these situations are rare, and for that reason, we neglect them in our approximation.

In third place, as discussed in Sect. 2.1, the spread's stationarity is a desirable property to look for in the selected pairs. However, a mean-reverting spread by itself does not necessarily generate profits. There must be coherence between the mean-reversion duration and the trading period. If a spread takes on average 400 days to mean-revert, but the trading period lasts one year, it is unlikely that a profitable situation is encountered. Likewise, a very short mean-reversion period is not desirable either. Since the half-life can be interpreted as an estimate of the expected time for the spread's mean-reversion, we propose filtering out pairs for which the half-life is not coherent with the trading period.

Lastly, we enforce that every spread crosses its mean at least once per month, to provide enough liquidity. It should be noted that even though there is logically a (negative) correlation between the number of mean crosses and the half-life period, as more mean crosses are naturally associated with a shorter half-life, both properties are not necessarily interchangeable. Adding this constraint might not merely enforce the previous condition but also discard pairs that in spite of meeting the mean-reversion timing conditions, do not cross the mean, providing no opportunities to exit a position.

Having described the four stages proposed to select an eligible pair, we proceed to describe the parameters used at each stage. Figure 3.5 represents the four rules a pair must verify to be eligible for trading. First, it is imposed that pairs are cointegrated, using a p-value of 1%. Then, the spread's Hurst exponent, represented by H should be smaller than 0.5. Additionally, the half-life period, represented by hl, should lay

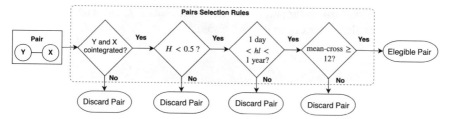

Fig. 3.5 Pairs selection rules

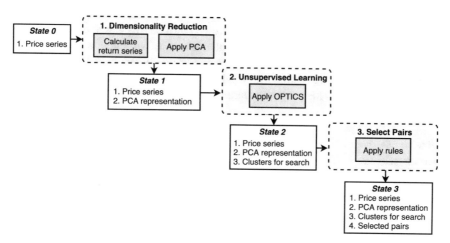

Fig. 3.6 Pairs selection diagram

between one day and one year. Finally, it is imposed that the spread crosses a mean at least 12 times per year, which ideally corresponds to a minimum of one cross per month, on average.

3.6 Framework Diagram

Up to this point, we described in detail the three building blocks of the proposed pairs selection framework. In this section, we intend to illustrate how they connect.

Figure 3.6 illustrates the progress along the three pairs stages. As we can observe, the initial state should comprise the price series for all the possible pairs' constituents. We assume this information is available to the investor. Then, by reducing the data dimensionality, each security may be described not just by its price series but also by the compact representation emerging from the application of PCA in the return series (State 1). Using this simplified representation, the OPTICS algorithm is capable of organizing the securities into clusters (State 2).

Finally, we may search for pair combinations within the clusters and select those that verify the rules described in Fig. 3.5 (State 3).

3.7 Conclusion

In this chapter, we started by motivating the exploration a novel way of grouping financial securities to search for pairs. Next, we proposed a compact data-driven scheme to accomplish that , and explored each of its stages. We described how the application of dimensionality reduction followed by a density-based clustering algorithm could help to tackle the previously stated issues. In particular, we presented the OPTICS method as being convenient for the conditions under study. We also defined a meticulous set of rules to define which pairs are eligible for trading. Some important take-away points from the proposed framework are summarized in Table 3.1.

Table 3.1 Key points from the proposed pairs selection framework

Proposed pairs selecion framework	
1. Pairs Search	Reasoning
Clustering using Unsupervised Learning	1.1. Data-driven, no bias from the investor;
	1.2. Tackles the multiple comparison problem;
	1.3. Alternative to the common grouping by sector approach;
Application of Principal Component Analysis	1.4. Extracts common underlying risk factors from securities' returns;
	1.5. Produces a compact representation for each security;
Application of OPTICS	1.6. No need to specify the number of clusters in advance;
	1.7. Robust to outliers;
	1.8. Suitable for clusters with varying density
2. Pairs Selection Criteria	Reasoning
Cointegrated pairs	2.1. Finds sound equilibrium relationships;
	2.2. The literature suggests cointegration performs better, when compared with minimum distance and correlation approaches;
Mean-reverting Hurst exponent	2.3. Provides an extra layer of confidence to validate mean-reverting series;
Suitable half-life	2.4. Filters out series with timings not compatible with the trading period;
Monthly mean crossing	2.5. Filters out pairs for which the spread does not converge to its mean with enough regularity (less than once per month)

References

1. Ankerst M, Breunig MM, Kriegel HP, Sander J (1999) Optics: ordering points to identify the clustering structure. ACM Sigmod Rec 28:49–60. ACM
2. Armstrong JS (2001) Principles of forecasting: a handbook for researchers and practitioners, vol 30. Springer Science & Business Media, Berlin
3. Avellaneda M, Lee JH (2010) Statistical arbitrage in the us equities market. Quant Financ 10(7):761–782
4. Bellman R (1966) Dynamic programming. Science 153(3731):34–37
5. Berkhin P (2006) A survey of clustering data mining techniques. In: Grouping multidimensional data. Springer, pp 25–71
6. Chan E (2013) Algorithmic trading: winning strategies and their rationale, vol 625. Wiley, New York
7. Ester M, Kriegel HP, Sander J, Xu X (1996) A density-based algorithm for discovering clusters in large spatial databases with noise
8. Harlacher M (2016) Cointegration based algorithmic pairs trading. PhD thesis, University of St. Gallen
9. Jolliffe I (2011) Principal component analysis. Springer, Berlin
10. MacQueen J et al (1967) Some methods for classification and analysis of multivariate observations. In: Proceedings of the fifth Berkeley symposium on mathematical statistics and probability, vol 1, Oakland, CA, USA, pp 281–297
11. Ramos-Requena J, Trinidad-Segovia J, Sánchez-Granero M (2017) Introducing hurst exponent in pair trading. Phys A: Stat Mech Appl 488:39–45
12. Ross SA (2013) The arbitrage theory of capital asset pricing. In: Handbook of the fundamentals of financial decision making: Part I. World Scientific, Singapore, pp 11–30
13. Rousseeuw PJ (1987) Silhouettes: a graphical aid to the interpretation and validation of cluster analysis. J Comput Appl Math 20:53–65

Chapter 4
Proposed Trading Model

4.1 Problem Statement

One downside of the threshold-based trading model, defined in Sect. 2.3.1, is that the entry points are not defined with much precision. The only criterion to enter a position is crossing a predefined threshold, independently of the spread's current direction. This can result in unpleasant portfolio decline periods, in case a pair continues to diverge, as depicted in Fig. 4.1.

The solid red line in Fig. 4.1 corresponds to the entry point defined by the model, which is triggered when the spread crosses the short threshold. Because the spread continues to diverge, a loss is incurred in the portfolio. Gains will only manifest once the spread starts converging, in the point marked by the dashed red line, at the end of February. In addition, entering the position at the point when convergence begins provides a larger profit margin since the spread distance to the mean is bigger. Hence, not optimizing the entry points does not only cause longer portfolio decline periods, but it can also reduce the full potential one could get from the spread convergence.

4.2 Proposed Model

To create a more robust trading model, it would be interesting to make use of predicted future data when defining the market entry positions. Initially, we conceive three different solutions based on time series forecasting. From these three approaches, which we proceed to describe, we select the most suitable based on empirical testing.

One possible approach could be the one followed by Dunis et al. [5] in Dunis et al. [3–5]. This approach attempts to model the spread predicted returns directly, which the author defines as

$$R_t = \frac{Y_t - Y_{t-1}}{Y_{t-1}} - \frac{X_t - X_{t-1}}{X_{t-1}}, \qquad (4.1)$$

© The Author(s), under exclusive license to Springer Nature Switzerland AG 2021
S. Moraes Sarmento and N. Horta, *A Machine Learning based Pairs Trading Investment Strategy*, SpringerBriefs in Computational Intelligence,
https://doi.org/10.1007/978-3-030-47251-1_4

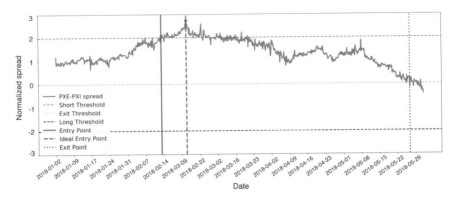

Fig. 4.1 Illustration of an untimely entry position

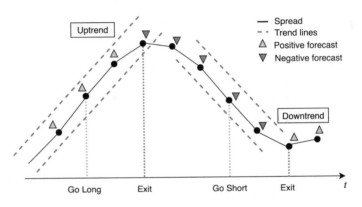

Fig. 4.2 Local momentum forecasting-based strategy

where X_t and Y_t represent the prices of the two legs composing the pair, at time t. This way, when the predicted return is higher than a predefined threshold (optimized in-sample), a position is entered.

Another possibility is to focus on modelling the spread series instead and use this information for trading. The investor may track the spread's recent values and look for a pattern of local trendiness. In case an uptrend is detected, the investor can go long on the spread. Similarly, if a downtrend is identified, the investor can go short on the spread. Eventually, a position is exited once the predicted value for the next time instant contradicts the trend's direction. This strategy is exemplified in Fig. 4.2. The coloured triangles express the forecasting input, used to indicate whether the spread is expected to go up (positive forecast) or down (negative forecast). For the sake of illustration, the forecasting has an accuracy of 100%, enabling the positions to be exited at the exact right moment. Naturally, this is extremely hard to achieve in practice.

Both methodologies described, even if theoretically appealing, did not perform well in practice. In Sect. 6.3.6, some details concerning why these two methodologies did not succeed are presented. From this point on, we focus on the approach that proves more reliable. This approach also focuses on forecasting the spread series, defined according to Eq. (2.11). In this case, it continuously monitors the percentage change between the spread at the current time, and the predicted spread in the next time-step. When the absolute value of the predicted change is larger than a predefined threshold, a position is entered, on the expectation that the spread suffers an abrupt movement from which the investor can benefit from. We proceed to describe the workings of this methodology more formally. We define S_t and S_t^* as the true and predicted value for the spread at time t, respectively. The predicted change is then given by Eq. (4.2a). Having calculated the spread predicted change, the position settlement conditions work as described in Eq. (4.2b), where α_L and α_S correspond to the long and short position trigger thresholds respectively. Once a position is entered, it is held while the predicted spread direction persists and closed when it shifts.

$$\text{Predicted Change} : \Delta_{t+1} = \frac{S_{t+1}^* - S_t}{S_t} \times 100 \tag{4.2a}$$

$$\text{Market entry conditions} : \begin{cases} \text{if } \Delta_{t+1} \geq \alpha_L, & \text{open long position} \\ \text{if } \Delta_{t+1} \leq \alpha_S, & \text{open short position} \\ \text{otherwise}, & \text{remain outside market} \end{cases} \tag{4.2b}$$

It is yet to be described how the thresholds (α_L, α_S) should be defined. A possible approach could consist of framing an optimization problem and try to find the profit-maximizing values. However, this approach is rejected due to its risk of data-snooping and unnecessary added complexity. We propose a simpler, non-iterative, data-driven approach to define the thresholds. For each spread series, we start by obtaining $f(x)$, the spread percentage change distribution during the formation period. The spread percentage change at time t is defined as

$$x_t = \frac{S_t - S_{t-1}}{S_{t-1}} \times 100.$$

From the distribution $f(x)$, the set of negative (negative percentage changes) and positive (positive percentage changes) values are considered separately, originating the distributions $f^-(x)$ (illustrated in Fig. 4.3a) and $f^+(x)$ (illustrated in Fig. 4.3b), respectively.

Since the proposed model targets abrupt changes but also requires that they occur frequently enough, looking for the extreme quantiles seems adequate. As quantiles adapt to the spread volatility, the thresholds will ultimately be linked to the spread's properties. As Fig. 4.3 illustrates, we recommend selecting the top decile and quintile[1] from $f^+(x)$ as candidates for defining α_L and the bottom ones, from $f^-(x)$, for

---

[1] A quintile is any of the four values that divide the sorted data into five equal parts, so that each part represents $\frac{1}{5}$ of the sample or population. It takes any of the values $Q_X(0.20), Q_X(0.40), \ldots, Q_X(0.80)$.

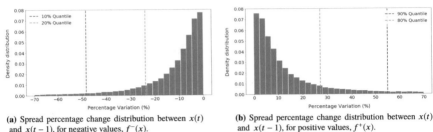

(a) Spread percentage change distribution between $x(t)$ and $x(t-1)$, for negative values, $f^-(x)$.

(b) Spread percentage change distribution between $x(t)$ and $x(t-1)$, for positive values, $f^+(x)$.

Fig. 4.3 Spread percentage change distributions

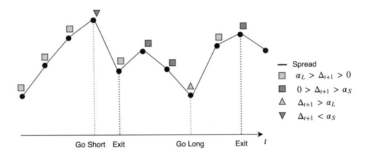

Fig. 4.4 Proposed forecasting-based strategy

defining α_S. Then, the quintile-based and decile-based thresholds are both tested in the validation set, and the most optimistic combination is adopted. The reason why we also consider quintiles and not just deciles is that we fear that the deciles might be too restrictive. The process just described can be presented more formally as

$$\{\alpha_S, \alpha_L\} = \underset{q}{\mathrm{argmax}}\; R^{\mathrm{val}}(q),\; q \in \left[\{Q_{f^-(x)}(0.20),\; Q_{f^+(x)}(0.80)\},\; \{Q_{f^-(x)}(0.10),\; Q_{f^+(x)}(0.90)\} \right]. \tag{4.3}$$

Figure 4.4 illustrates the application of this model. The colored symbols are used to express the information provided by the forecasting algorithm. The triangles indicate one of the thresholds has been triggered and the corresponding direction, whereas the squares only provide information concerning the predicted direction. Once again, for the sake of illustration, the forecasting has perfect accuracy, meaning the positions can be set in optimal conditions.

As a final note, it is worth emphasizing that the methodology just proposed is fundamentally different from the benchmark trading strategy presented in Sect. 2.3.1. Because the underlying motivation for entering a position is distinct in both models, it is not expected any relation between the entry and exit points of the two strategies.

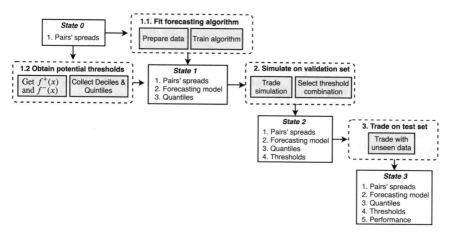

Fig. 4.5 Proposed model application diagram

4.3 Model Diagram

The proposed model can be applied in line with Fig. 4.5. As depicted, the investor should start by training the forecasting algorithms to predict the spreads. Furthermore, the decile-based and quintile-based thresholds should be collected to integrate the trading model. Having fitted the forecasting algorithms and obtained the two combinations for the thresholds (State 1), the model can be applied on the validation set. From the validation performance, the best threshold combination is selected (State 2). At this point, the model is finally ready to be applied on unseen data, from which the performance may be inferred (State 3).

4.4 Time Series Forecasting

Motivated by the potential of applying a time series forecasting based trading strategy, we proceed to describe which alternatives may be more suitable for the task at hand. Different models have been applied in the literature for time series forecasting. These models can be divided into two major classes, parametric and non-parametric models.

Parametric models appear as the early time series predictions and assume that the underlying process has a particular structure, which can be described using a small number of parameters. In these approaches, the goal is to estimate the parameters of the model that best describe the stochastic process. However, these models' representation capacity have severe limitations Chen et al. [2], creating the need to search for a modelling approach with no structural assumptions.

Nonparametric modelling is especially attractive because it makes no structural assumptions about the underlying structure of the process. Artificial Neural Networks

(ANN) are an example of non-parametric models. Several factors motivate their use. First, ANNs have been an object of attention in many different fields, which makes its application in this context an interesting case study. Furthermore, ANN-based models have shown very promising results in predicting financial time series data in general Cavalcante et al. [1]. Finally, Machine Learning techniques are fairly unexplored in this field, as analyzed in Sect. 2.4. This encourages its application in this work.

In conclusion, we propose the application of a benchmark parametric approach, along with other non-parametric models found relevant. Each of the proposed algorithms is detailed next.

4.5 Autoregressive Moving Average

As stated by Si and Yin [9], financial time series are inherently complex, highly noisy and nonlinear. These aspects naturally motivate the use of ANNs for forecasting, given their capability to address such conditions. However, the financial time series under analysis do not necessarily fit the characterization above. By construction, the spread series is stationary during the formation period. This is a requirement, given that a pair is only selected if both constituents are cointegrated, which consequently implies that its spread is stationary. Therefore, even though stationarity in the trading period cannot be guaranteed, because it corresponds to unseen data, the series is surely more likely to be so. It is thus fair to ask if a simpler model could perform this task just as effectively. Moreover, it would be interesting to determine whether the extra model complexity from neural networks is worthwhile. The application of an autoregressive moving-average (ARMA) model for forecasting is proposed. The use of an ARMA model also allows inferring to what extent the profitability of the strategy depends on the time series forecasting algorithm itself, or if it is robust under different forecasting conditions.

The ARMA model was first described by Whitle [10]. This model describes a stationary stochastic process as the composition of two polynomials. The first polynomial, the autoregression (AR), intends to regress the variable at time t on its own lagged values. The second polynomial, the moving average (MA), models the prediction error as a linear combination of lagged error terms and the time series expected value. We proceed to describe in a formal way how each of the polynomials models a time series X_t. The AR(p) model is described as

$$X_t = c + \sum_{i=1}^{p} \varphi_i X_{t-i} + \varepsilon_t, \tag{4.4}$$

where p is the polynomial order, $\varphi_1, \ldots, \varphi_p$ are the model parameters, c is a constant, and ε_t is a random variable representing white noise. The MA(q) model is described as

$$X_t = \mu + \varepsilon_t + \sum_{i=1}^{q} \theta_i \varepsilon_{t-i}, \qquad (4.5)$$

where q is the polynomial order, $\theta_1, \ldots, \theta_q$ represent the model parameters and μ represents the mean value of X_t. The variables $\varepsilon_t, \varepsilon_{t-1}, \ldots, \varepsilon_{t-q}$ correspond to white noise error terms in the corresponding time instants. Finally, the ARMA(p,q) model results from the combination of (4.4) and (4.5), and can be represented as

$$X_t = c + \varepsilon_t + \sum_{i=1}^{p} \varphi_i X_{t-i} + \sum_{i=1}^{q} \theta_i \varepsilon_{t-i}. \qquad (4.6)$$

4.6 Artificial Neural Network Models

Although the simplicity of the ARMA model just described is advantageous from a practical point of view, it might also be interpreted as a limitation for learning more complicated data patterns. Artificial Neural Networks have proved particularly efficient in this regard. From the vast amount of existing ANNs configurations, we proceed to select the two architectures deemed more appropriate for this work.

4.6.1 Long Short-Term Memory

One configuration considered suitable is the Long-Short Term Memory (LSTM). The LSTM is a type of Recurrent Neural Network (RNN). Therefore, RNNs should first be described. Unlike standard feedforward neural networks (FNN), RNNs can process not only single data points but also entire sequences of data Goodfellow et al. [7]. This means that while in a traditional FNN, inputs are assumed to be independent of each other, in the case of an RNN, a sequential dependency is assumed to exist among inputs, and previous states might affect the decision of the neural network at a different point in time, as represented in Fig. 4.6. This is achieved with inner loops, allowing RNNs to have a memory of their previous computations.

In Fig. 4.6, W_i and b_i represent the neural networks weights, and the bias term respectively. The subscript i can take the values of x for the input, h for hidden and o for output. The hidden layer's RNN cell, represented by the grey rectangle, is a block which repeats (unfolds) along the time dimension. This block receives an input and together with the previous state computes the next state and the output at the current time step. The two cell outputs can be described as

$$\begin{cases} h_t = \sigma \left(W_h h_{t-1} + W_x x_t + b_h \right) \\ o_t = \sigma \left(W_o h_t + + b_o \right) \end{cases}, \qquad (4.7)$$

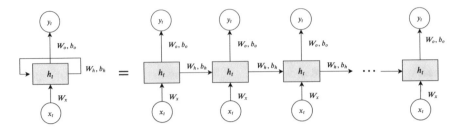

Fig. 4.6 Recurrent Neural Network structure

where σ represents the layer's activation function. A problem arising with the RNN architecture just presented is the difficulty in propagating the prediction error. To understand why we first describe the standard backpropagation in regular FNNs. In that case, the backpropagation moves backwards from the final error through the outputs, weights, and inputs of each hidden layer, considering those weights accountable for a portion of the error, by calculating their partial derivatives $\frac{\partial E}{\partial w_i}$, where E represents the error function. Those derivatives are then used to adjust the weights in whichever direction decreases error. To train the RNN models, the same process applies, by representing the network on its unfolded version, as in Fig. 4.6. However, when propagating the error along the entire network's time dimension, the gradient might vanish or explode, depending on whether the derivative of the state update function decreases or increases the backpropagation error too drastically. In the first case, the learning process takes too long to complete, whereas the second case leads to numerical issues.

The aspect described is partly mitigated with LSTMs Hochreiter and Schmidhuber [8] by selectively updating the internal state as a function of the current state and the input, through a gating mechanism. In this topology, the internal state and current input define the input, output and forget gate's flow of information. The input gate defines the influence of the new input on the internal state. The forget gate defines which parts of the internal state are retained over time. Finally, the output gate defines which part of the internal state affects the output of the network. Formally, the equations describing the LSTM cell can be summarized as

$$\begin{cases} f_t = \sigma \left(W_x^f x_t + W_h^f h_{t-1} + b_f \right) \\ i_t = \sigma \left(W_x^i x_t + W_h^i h_{t-1} + b_i \right) \\ o_t = \sigma \left(W_x^o x_t + W_h^o h_{t-1} + b_o \right) \\ c_t = f_t \cdot c_{t-1} + i_t \cdot \sigma \left(W_x^c x_t + W_h^c h_{t-1} + b_c \right) \\ h_t = o_t \cdot \sigma \left(c_t \right) \end{cases} , \qquad (4.8)$$

where f_t, i_t and o_t represent the forget, input and output gate respectively. By selectively accepting new information and erasing past one, this topology is able to maintain the error flow across the network.

4.6.2 LSTM Encoder-Decoder

From a trading perspective, it might be beneficial to use information regarding the prediction not just of the next time instant, but also of later time steps. To accomplish that, one might use the data collected up to time t to directly predict the next N time steps $(t + 1, \ldots, t + N)$ at once. An improvement may consist of using not just the data collected up to time t, but also benefit from the new predicted data. For instance, to predict the time series value at time $t + 2$, one would use the data collected up to t with the addition of the prediction for $t + 1$.

An LSTM Encoder-Decoder architecture is naturally fitted to such scenarios. This architecture is composed by two groups of LSTMs, one for reading the input sequence and encoding it into a fixed-length vector, the encoder, and a second for decoding the fixed-length vector and outputting the predicted sequence, the decoder. The two networks are trained jointly to minimize the forecasting loss with respect to a target sequence given the source sequence. The architecture of a LSTM based Encoder-Decoder can be observed in Fig. 4.7. A sequence $x_1, \ldots, x_t$ is inputted sequentially to the Encoder LSTM, which aggregates the relevant information in the encoder vector. Then, this information is used to forecast the series for the next N time-steps. Naturally, it is expected that the error increases with N.

In this multi-step forecasting scenario, the trading rules defined in Sect. 4.2 need to be updated. The prediction change should be calculated N times in advance. Likewise, the thresholds α_L and α_S should be calculated with respect to the distribution of the percentage change between $x(t)$ and $x(t - N)$.

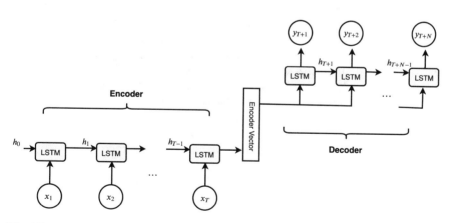

Fig. 4.7 LSTM Encoder-Decoder

4.7 Artificial Neural Networks Design

The Artificial Neural Network models demand a laborious design process. In this section, we highlight some techniques adopted, which we consider especially relevant to describe.

4.7.1 Hyperparameter Optimization

One of the disadvantages pointed out to Deep Learning models is the cumbersome process associated with optimizing the network's configuration. Tuning the hyper-parameters can have a tremendous influence on its result, but it is also very time-consuming. Thus, a compromise must be achieved.

Due to the limited computation resources, the tuning of the LSTM models is constrained to a set of most relevant variables: (i) sequence length, (ii) number of hidden layers, and (iii) nodes in each hidden layer. We experiment increasingly complex configurations until signs of overfitting are detected.

4.7.2 Weight Initialization

Neural networks weights are very often initialized randomly. This initialization may take a fair amount of iterations to converge to the optimal weights. Moreover, this kind of initialization is prone to vanishing or exploding gradient problems. One way to reduce this problem is to select a random weight initialization carefully. In this work, we adopt the glorot weight initialization, proposed by Glorot and Bengio [6]. The authors introduce a weight initialization algorithm which factors into the equation the size of the network (number of input and output neurons). The proposed methodology does not only reduce the chances of running into gradient problems but also brings substantially faster convergence.

This initialization, also called Xavier initialization, suggests sampling the layer weights in a way that preserves the input variance, such that the variance remains constant as the information passes through each layer of the neural network. We proceed to describe how this can be accomplished. First, the variance is given by

$$Y = \sum_{N_{in}} W_i X_i \Leftrightarrow \text{Var}(Y) = \text{Var}\left(\sum_{N_{in}} W_i X_i\right), \qquad (4.9)$$

where N_{in} represents the number of input nodes. If the input variables are assumed to be uncorrelated, we may consider

$$\text{Var} \left(\sum_{N_{in}} W_i X_i \right) = \sum_{N_{in}} \text{Var} \left(W_i X_i \right). \tag{4.10}$$

Finally, assuming the inputs and weights are uncorrelated and that the variance of both inputs and weights is identically distributed, (4.10) further simplifies according to

$$\sum_{N_{in}} \text{Var} \left(W_i X_i \right) = \sum_{N_{in}} \text{Var} \left(W_i \right) \text{Var} \left(X_i \right) = N_{in} \left(\text{Var} \left(X_i \right) \text{Var} \left(W_i \right) \right). \tag{4.11}$$

Thus, to preserve the input variance, $\text{Var}(X_i)$, the weights variance should be given by $\frac{1}{N_{in}}$. To accomplish this, a Gaussian sampling distribution with zero mean and a variance given by $\frac{2}{N_{in}+N_{out}}$ is proposed. This corresponds to the geometric mean between the forward propagation pass variance, $\frac{1}{N_{in}}$, and back propagation pass variance, $\frac{1}{N_{out}}$.

4.7.3 Regularization Techniques

The functions that define neural network topologies are generally non-convex Goodfellow et al. [7]. Therefore, the training algorithm does not give us a formal guarantee of convergence to the optimal solution. The imposition of regularization terms in the optimization process aids tackling this issue. Regularization refers to the introduction of additional constraints, potentially derived from problem-specific information, to solve an ill-posed problem or avoid overfitting. More specifically, it helps to reduce the generalization error by making the gap between the errors observed in the training set and validation set less evident. At the same time, by introducing additional information, the convergence rate can increase Goodfellow et al. [7]. We proceed to describe two regularization techniques adopted in this work, Early-stopping and Dropout.

Early-stopping is likely the most commonly used form of regularization in Deep Learning due both to its effectiveness and its simplicity Goodfellow et al. [7]. The foundation of Early-stopping goes as follows. When training large models, with sufficient representational capacity to overfit, it is often visible that training error decreases steadily over time. However, the validation set error begins to rise again at some point. This behaviour is illustrated in Fig. 4.8. In this figure, the described pattern is particularly evident.

By accounting for this pattern, a model with a better validation set error (and hopefully a better test set error) can be obtained. To achieve that, we should return to the parameter setting with the lowest validation error, identified by the black triangle in Fig. 4.8. This can be accomplished by maintaining a copy in memory of the model parameters that so far generated the best validation error. When the training

Fig. 4.8 Early Stopping
illustration

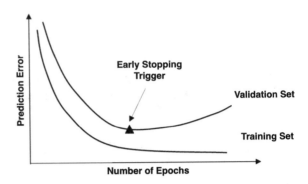

terminates, the algorithm returns the stored parameters, rather than the most recent ones. This process can be made more efficient if the system terminates the training procedure after detecting no improvement in the validation loss within a number of p iterations after the best parameterization, rather than having to wait for the training to terminate. The parameter p is usually called patience.

The second regularization technique adopted is Dropout. Its fundamental idea is based on the premise that the risk of overfitting may be reduced by fitting different neural networks to the same dataset and use the average prediction of all networks as the final prediction. The view is that noise learned by a specific model architecture is attenuated when averaged over all models. However, the process of training multiple models can be very challenging, especially when dealing with such complex models. Dropout emerges as a more practical solution to emulate this. It approximates the training of a large number of neural networks with different architectures in parallel, while only training one neural network. This is accomplished by randomly ignoring some layer nodes during the training period. This has the effect of treating the layers like they have a different number of nodes and connectivity to the prior layer. Hence, each update during training is performed with a different perspective of the configured layer.

4.8 Conclusion

In this chapter, we started by emphasizing the motivation for exploring a new trading strategy. We advanced with the description of the proposed trading model. Next, we explored the potential forecasting algorithms to integrate it. Namely, we stated that the application of an ARMA model, a regular LSTM neural network and an LSTM Encoder-Decoder seem adequate in this context. In addition, we discussed some technical aspects concerning the design of the ANN-based models. Table 4.1 compiles some key points concerning the proposed trading model.

Table 4.1 Key points from the proposed trading model

Proposed trading model	
1. Model guidelines	Reasoning
Forecast the pair's spread	1.1. Stationary time-series, easier to predict in theory;
	1.2. Returns proved harder to forecast;
Monitor predicted spread change	1.5. Predicted spread percentage change is a good indicator of upcoming deviations, and therefore good trading opportunities;
Quantile-based thresholds	1.6. Model parameters automatically adapt to the data;
	1.7. Simple to implement.
2. Forecasting algorithms	Reasoning
ARMA	2.1. Suited for stationary time-series;
	2.2. Benchmark to compare with more complex models;
	2.3. Infer trading model dependency on the forecasting algorithm;
LSTM	2.4. Demonstrated ability in the literature;
	2.5. Learns non-linear representations of the data;
	2.6. Can memorize long sequences;
LSTM Encoder Decoder	2.7. Naturally provides multi-step forecasts;
	2.8. Preserves the advantages from LSTM.

References

1. Cavalcante RC, Brasileiro RC, Souza VL, Nobrega JP, Oliveira AL (2016) Computational intelligence and financial markets: a survey and future directions. Expert Syst Appl 55:194–211
2. Chen G, Abraham B, Bennett GW (1997) Parametric and non-parametric modelling of time series' an empirical study. Environ: Off J Int Environ Soc 8(1):63–74
3. Dunis CL, Laws J, Evans B (2006) Modelling and trading the gasoline crack spread: a non-linear story. Deriv Use Trading Regul 12(1–2):126–145
4. Dunis CL, Laws J, Evans B (2009) Modelling and trading the soybean-oil crush spread with recurrent and higher order networks: a comparative analysis. In: Artificial higher order neural networks for economics and business, IGI Global, pp 348–366
5. Dunis CL, Laws J, Middleton PW, Karathanasopoulos A (2015) Trading and hedging the corn/ethanol crush spread using time-varying leverage and nonlinear models. Eur J Financ 21(4):352–375
6. Glorot X, Bengio Y (2010) Understanding the difficulty of training deep feedforward neural networks. In: Proceedings of the thirteenth international conference on artificial intelligence and statistics, pp 249–256
7. Goodfellow I, Bengio Y, Courville A (2016) Deep learning. MIT Press, Cambridge
8. Hochreiter S, Schmidhuber J (1997) Long short-term memory. Neural Comput 9(8):1735–1780
9. Si YW, Yin J (2013) Obst-based segmentation approach to financial time series. Eng Appl Artif Intell 26(10):2581–2596
10. Whitle P (1951) Hypothesis testing in time series analysis, vol 4. Almqvist & Wiksells, Stockholm

Chapter 5
Implementation

5.1 Research Design

In this work, we propose a methodology for each phase composing a Pairs Trading strategy, the pairs selection phase (Research Stage 1) and trading phase (Research Stage 2). Thus, we design a dedicated research setup for each, as illustrated in Fig. 5.1.

One convenient research scheme consists of responding to this work's first research question, and apply the optimal pairs clustering technique just found to answer the second research question. This is represented by the greyed arrow in between. The coloured blocks intend to facilitate the visualization of the Pairs Trading phase being tested.

5.2 Dataset

In this section, we present the dataset suggested for simulating the proposed implementation. We start by describing which type of securities we find appropriate to explore. Next, we describe the resulting dataset and how it is prepared for application. Finally, we detail how the data is partitioned in several periods, to be used in different experiments.

5.2.1 Exchange-Traded Funds

Exchange-Traded Funds (ETFs) are considered an interesting type of securities to explore in this work. An ETF is a security that tracks an index, a commodity or a basket of assets just like an index fund, but trades like a stock. According to Chan [4], the one advantage of trading ETF pairs instead of stock pairs is that once found to be cointegrated, ETF pairs are less likely to fall apart in out-of-sample data. This is because the fundamental economics of a basket of stocks changes much slower than that of a single stock. This robustness is very appealing in the context of

© The Author(s), under exclusive license to Springer Nature Switzerland AG 2021
S. Moraes Sarmento and N. Horta, *A Machine Learning based Pairs Trading Investment Strategy*, SpringerBriefs in Computational Intelligence, https://doi.org/10.1007/978-3-030-47251-1_5

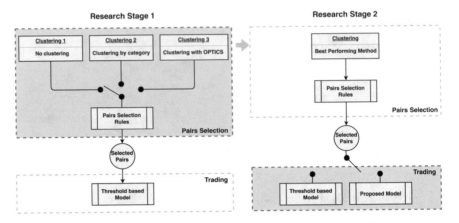

Fig. 5.1 Research design overview

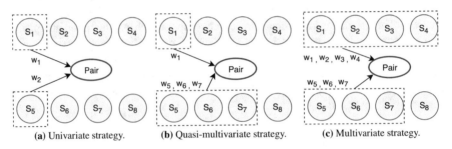

Fig. 5.2 Exemplification of Univariate, Quasi-Multivariate and Multivariate strategies

Pairs Trading, in which the fundamental premise is that securities whose prices are cointegrated today should remain cointegrated in the future.

To further support the hypothesis that ETFs are convenient in a Pairs Trading setting, the work of Chen et al. [5], and Perlin [18] are an important reference. To better comprehend the authors' findings, some nomenclature should be introduced first. A strategy in which both constituents of the pair are single securities is called univariate. Whereas if one side of the pair is formed by a weighted average of comoving securities, the strategy is named quasi-multivariate. Reasonably, when both sides of the spread are formed by a weighted average of a group of securities, the strategy is said to be multivariate, as represented in Fig. 5.2.

Proceeding with the analysis of the aforementioned work, Chen et al. [5], and Perlin [18] attempt to produce more robust mean-reverting time series by using a linear combination of stocks to form each leg of the spread, on the prospect that one outlier stock will not affect the spread so drastically. In specific, Chen et al. [5] applies a Pairs Trading strategy based on Pearson correlation both in a univariate and a quasi-multivariate setup. In the former, the two most correlated stocks form a pair. In the latter, the pair is formed by a stock and an equal-weighted portfolio of the 50 most correlated stocks. Results show that the quasi-multivariate trading strategy obtains a better performance, whereas reducing the number of stocks in the weighted portfolio causes the returns to drop up to two thirds. Perlin [18] also evaluates the efficiency of a univariate strategy against a quasi-multivariate strategy, for the 57 most liquid stocks in the Brazilian market. In the univariate scenario, the two most correlated stocks are identified as a pair, whereas in the quasi-multivariate scenario, one stock is paired with five partner stocks. The results obtained show that the quasi-multivariate strategy outperforms the univariate strategy, once again. Nevertheless, both studies mentioned in this paragraph see their results affected by the high associated transaction costs of purchasing multiple securities, which disincentive the application of a multivariate strategy. In this work, we suggest using ETFs as a proxy to obtain the benefits of multivariate strategies more practically. The reasoning is that an ETF is fundamentally a weighted average of a set of securities, and thus, trading ETFs may be seen as a more practical way of running a multivariate pairs trading strategy.

On top of the motivations described, the universe of ETFs has one extremely interesting particularity. The rising popularity of ETFs motivates the creation of new ETFs every year that just marginally differ from existing ones. These price series naturally form good potential pairs.

Finally, few research studies explore the application of Pairs Trading in the ETF universe, with some exceptions Avellaneda and Lee [1], Galenko et al. [11], which makes it even more appealing to try the suggested approaches using ETFs.

5.2.2 Data Description

This work focuses on a subset of ETFs that track single commodities, commodity-linked indexes or companies focused on exploring a commodity. Constraining the ETFs to this subset reduces the number of possible pairs, making the strategy computationally faster and leaving space for a careful analysis of the selected pairs. More specifically, the set of ETFs considered is composed only by commodity-linked ETFs that are available for trading in January 2019. To find this information, the data source ETF.com [10] proved useful.

A total of 208 different ETFs verify the conditions just described. This universe of commodity-linked ETFs can be categorized according to the type of commodities being tracked. We propose five different categories, as shown in Table 5.1. Appendix 1 includes a table describing all the eligible ETFs.

It should be noted that by considering just the ETFs active throughout an entire period, survivorship bias is induced. By neglecting delisted ETFs, possible positions that would have had to be left abruptly are not being accounted for in the results. This is because no survivorship free dataset containing the pretended ETFs was found. To attenuate this impact, recent periods are considered whenever possible.

Table 5.1 Commodity categories considered in the dataset

Category	Description
Agriculture	ETFs tracking agricultural commodities, including staple crops and animals produced or raised on farms or plantations
Energy	ETFs chasing energy commodities, which are mostly hard commodities that are mined or extracted. It includes fossil fuels like Coal, Oil and Natural gas
Base metals	ETFs focused on base metals which support a whole range of industrial and commercial applications including construction and manufacturing
Precious metals	ETFs monitoring precious metals, naturally occurring metallic elements with high economic value
Broad market	ETFs following the movement of the commodity market as a whole, rather than a specific sector

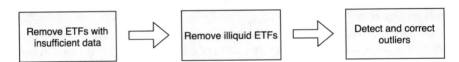

Fig. 5.3 Data preprocessing steps

Having selected the universe of ETFs to consider for trading, we proceeded to retrieve the price series for each ETF. We considered price series data with 5-min frequency. The motivation for using intraday data is three-fold. First, with finer granularity, the entry and exit points can be defined with more precision, leaving space for higher profits margins. Secondly, we may detect intraday mean-reversion patterns that could not be found otherwise. Lastly, it provides more data samples, allowing to train complex forecasting models with less risk of overfitting.

5.2.3   Data Preparation

Before delving into the search of promising pairs, some data cleaning steps are implemented, as described in Fig. 5.3.

First, ETFs with any missing values are directly discarded. Then, we remove ETFs that do not verify the minimum liquidity requisites, to ensure the considered transaction costs associated with bid-ask spread are realistic.[1] The minimum liquidity requisites follow the criterion adopted in Do and Faff [7], Gatev et al. [12], which discards all ETFs not traded during at least one day.

At last, some price series contain occasional outliers. In the literature, several techniques have been proposed to tackle the issue of anomaly detection. Resorting

---

[1] Trading illiquid ETFs would result in higher bid-ask spreads which could dramatically impact the profit margins.

to methodologies such as cluster analysis based techniques, Hidden Markov Models or other Machine Learning techniques could be considered for this purpose. Nonetheless, because these events are infrequent in the considered dataset, a simpler technique is proposed. For each price series, the return series is calculated. Then, every point for which the percentage change is higher than 10%, which is extremely unlikely to occur in just 5 min, is considered an outlier. These situations are analyzed one-by-one. In case there is a single outlier in the time series of an ETF, this point is corrected manually by analyzing other data sources. When more outliers are present, the ETF is discarded.

Additionally, when training the forecasting-based models, the data is standardized according to

$$\overline{S}_t = \frac{S_t - \mu_s}{\sigma_s}, \tag{5.1}$$

where $\overline{S}_t$ represents the result of standardizing the spread S_t. The resulting series has zero mean and unit variance. Standardizing the inputs can make the training of the neural networks faster and reduce the chances of getting stuck in local optima.

5.2.4 Data Partition

For each trading simulation, the data must be partitioned in two periods. The formation period and the trading period. The formation period simulates the data available to the investor before engaging in any trade, and its role is three-fold. First, this period is used to find the most appealing candidate pairs. Secondly, a smaller portion of the data, the validation set, is used to simulate how the strategy would have performed in recent times. Thirdly, and only applicable to the forecasting-based models, a portion of the formation period data, the training set, is used to train the forecasting algorithm.[2] The trading period is used to resemble how the implemented trading model would perform with future unseen data. Figure 5.4 illustrates how this periods are arranged.

It is enticing to consider several partitions of the dataset, since analyzing the trading results for different conditions provides more confidence in the statistical significance of the results obtained. The typical procedure in classical Machine Learning problems is to perform k-fold cross-validation. However, time series (or other intrinsically ordered data) can be problematic for cross-validation, as one wants to avoid the look-ahead bias that may arise from training the model with data not available at a given historical date. To tackle this issue, a moving training and test window are used sequentially. Forward chaining could also be applied, but in that case, the training and test periods' size is not fixed. Hence, the results can be affected by the varying data size.

---

[2]To be more precise, the training set is used to train the ANN-based models, whereas the ARMA model is trained on the entire formation period, as described in Sect. 6.3.2.

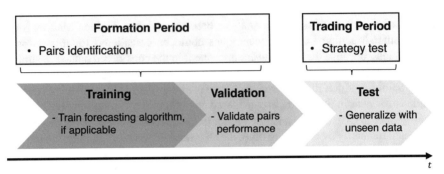

Fig. 5.4 Period decomposition

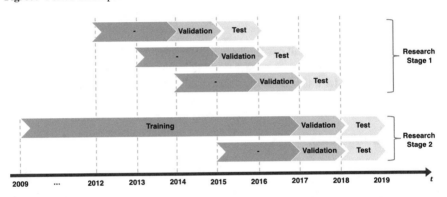

Fig. 5.5 Data partition periods

The periods considered for simulating each research stage are illustrated in Fig. 5.5. There are essentially two possible configurations: (i) the 3-year-long formation periods, and (ii) the 9-year-long formation period. In both cases, the second-to-last year is used for validating the performance, before running the strategy on the test set. We define a 1-year-long trading period, based on the findings of Do and Faff [6], that claim the profitability can be increased if the initial 6-month trading period proposed in Gatev et al. [12] is extended to 1 year.

The 3-year-long formation period is adopted when using the threshold-based trading model. Although it is slightly longer than what is commonly found in the literature,[3] we decide to proceed on the basis that a longer period may identify more robust pairs. The 9-year-long formation period is used for simulating the forecasting-based trading model. In this case, more formation data is required to fit the forecasting algorithms. The first 8 years are used for training them, as indicated in Fig. 5.5.

---

[3]Do and Faff [6], Gatev et al. [12], Rad et al. [19] use a 1-year-long formation period. Dunis et al. [9] makes use of a 3-month formation period.

In this case, the application of several test periods is not conceivable due to the computational burden of training the associated forecasting algorithms. The number of neural networks requiring training when using T periods can be calculated as

$$\sum_{i=1}^{T} M_i \times N_i, \tag{5.2}$$

where M_i is the number of pairs found in the i-th formation period, and N_i is the number of model configurations to be tested in the i-th period (e.g. RNN, Encoder-Decoder, etc.). Therefore, studying several periods consumes a considerable amount of extra time. Because we want to take special attention to the configuration of each model, we find more prudent to use a single formation/test period,[4] and utilize the available time to optimize the Deep Learning architectures and corresponding parameters meticulously.

One question that naturally arises is whether the threshold-based model used to compare with the forecasting-based model, in research stage 2 (see Fig. 5.1), should be tested under the same period (2009–2018) or in the conditions previously applied (3-year-long formation period). We suggest conducting both scenarios to distinguish which part of the results is explained by the trading model itself and what part is affected by the period duration. For this reason, the period ranging from 2015 to 2019 is selected to test the benchmark model.

Table 5.2 describes some relevant data concerning the data partitions described in this section. The majority of ETFs found are relatively recent. Consequently, when longer periods are considered, the number of active ETFs drops significantly.

5.3   Research Stage 1

This section describes how we propose to evaluate the pairs selection framework introduced as part of research stage 1. We suggest comparing the three pairs' search methods mentioned in this work: (i) searching for all combinations of pairs, (ii) searching for pairs within the same category and (iii) searching for pairs within clusters generated using OPTICS. Therefore, we proceed to detail the necessary steps to build this comparison setup.

---

[4]Although it is not mentioned explicitly, we suspect the computational effort is also one motivation that led Dunis et al. [8], Krauss et al. [16] to consider a single displacement for training, validation, and test sets.

Table 5.2 Dataset partitions (sampled with a frequency of 5 min)

Research stage	Research stage 1			Research stage 2	
Period	2012–2015	2013–2016	2014–2017	2009–2018	2015–2018
Trading days	1 004	1 007	1 007	2 515	1 006
Begin	03-01-2012	02-01-2013	02-01-2014	02-01-2009	02-01-2015
End	30-12-2015	30-12-2016	29-12-2017	28-12-2018	31-12-2018
Samples (5 min)	77 916	78 222	78 258	195 414	78 468
Train	38 940	39 096	39 132	156 516	39 312
Validation	19 470	19 506	19 620	19 506	19 578
Test	19 506	19 620	19 506	19 392	19 578
Number of ETFs	95	105	116	58	125
Agriculture	11	11	11	7	11
Broad market	11	11	12	7	12
Energy	42	48	54	29	58
Industrial metals	5	6	6	2	7
Precious metals	26	29	33	13	37

5.3.1 Development of the Pairs Selection Techniques

To compare each of the three different pairs' search techniques, each approach is built from scratch. We also develop all the pairs selection rules proposed in Sect. 3.5.

We find relevant to emphasize that grouping by category requires establishing a mapping between every ETF and the corresponding group. To accomplish this, we start by retrieving the available information regarding the segment of each ETF, found in ETF.com [10]. The segment represents the precise market to which the fund provides exposure to. From this information, a second mapping is constructed containing the correspondence between every segment and one of the five categories described in Table 5.1.

5.3.2 Trading Setup

Since the main goal of this research stage is to compare the results from the pairs selection techniques relative to each other, we do not worry about optimizing the trading model. Therefore, the standard threshold-based model described in Sect. 2.3.1 is applied, with the parameters specified in Table 5.3.

Table 5.3 Threshold-based model parameters

Parameters	Values
Long treshold	$\mu_s - 2\sigma_s$
Short threshold	$\mu_s + 2\sigma_s$
Exit threshold	μ_s

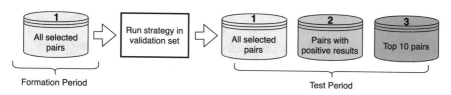

Fig. 5.6 Test portfolios

The model parameters used are the ones suggested by Gatev et al. [12], which serve as a reference to several studies in the field. The spread's standard deviation, σ_s, and mean, μ_s, are calculated with respect to the entire formation period. The exit threshold is defined as the mean value, meaning that a position is closed only when the spread reverts to its mean.[5] We are aware that there exists some research work which explicitly explores the optimization of the trading thresholds, such as Göncü and Akyıldırım [13], Huang et al. [14]. Also, some authors, as Dunis et al. [9], explore the application of sliding windows to constantly update μ_s and σ_s. Nevertheless, we stick to this framework for simplicity.

5.3.3 Test Portfolios

We implement three different test portfolios resembling probable trading scenarios. Portfolio 1 considers all the pairs identified in the formation period. Portfolio 2 takes advantage of the feedback collected from running the strategy in the validation set by selecting only the pairs that had a positive outcome. Lastly, Portfolio 3 corresponds to the situation in which the investor is constrained to invest in a fixed number of k pairs. In such a case, we suggest selecting the top-k pairs according to the returns obtained in the validation set. We consider $k = 10$, as it stands in between the choices of Gatev et al. [12], which uses $k = 5$ and $k = 20$. By testing different portfolio constructions, we may not only find the optimal clustering procedure but also evaluate its best condition of application. The three portfolios are represented in Fig. 5.6.

---

[5]Note that the parameters are simplified when working with the standardized spread, because in that case $\mu_s = 0$ and $\sigma_s = 1$.

It is worth emphasizing that even though selecting the pairs based on the performance on the validation set seems appropriate, there is no guarantee that the best performing pairs in the validation set will follow the same behaviour in the test set. The results will dictate the competence of this heuristic.

5.4   Research Stage 2

In this section, we describe the proposed implementation to gauge the potential of the forecasting-based trading model introduced, as part of research stage 2.

5.4.1   Building the Forecasting Algorithms

To compare the robustness provided by the standard threshold-based model with the proposed forecasting-based model, simulated using the ARMA model, an LSTM and an LSTM Encoder-Decoder, the three models are built from scratch. Due to the limited available resources and the complexity of training these models, there are severe constraints on the number of combinations we can experiment. For this reason, the architecture of each model is tuned in a manual iterative process based on the feedback from the results obtained. The training is stopped when the results evidence signs of overfitting. Concerning the LSTM Encoder-Decoder architecture, we propose that it outputs two time-steps at a time. Forecasting the time series two steps in advance seems appropriate, as we fear that the precision deteriorates for more distant values.

A naive baseline defining a lower bound objective is set to support the training process. The persistent forecasting algorithm is selected for this purpose. It uses the value at the previous time step to predict the expected outcome at the next time step, according to

$$\text{Naive} : Y_{t+1} = Y_t. \tag{5.3}$$

5.4.2   Test Conditions

Comparing the forecasting-based models using the three different clustering techniques would be very costly. For this reason, as illustrated in Fig. 5.1, we consider the application of the clustering methodology that proves to be more appealing. As for the test portfolios, we consider the option of applying the strategy exclusively to the pairs that obtained positive results during the validation period (Portfolio 2 in Fig. 5.6). Relying on selecting the top 10 pairs also seemed an adequate choice. However, as we will see, it might be the case that the selected pairs are fewer than ten.

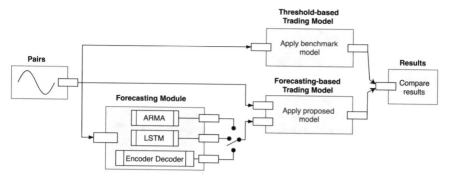

Fig. 5.7 Trading models interaction diagram

Figure 5.7 summarizes how the proposed methodology will be set aside with the threshold-based model in the simulation covering the period of January 2009 to December 2018.

5.5 Trading Simulation

In this section, we describe the details regarding the trading simulation. We go over the aspects considered in the construction of the portfolio, the transactions costs, and the position settlement conditions.

5.5.1 Portfolio Construction

We start by noting that the portfolios studied in this work may have varying sizes. Nevertheless, we establish that all pairs should be equally weighted in the portfolio. This way, the portfolio returns can be obtained by simply averaging the performance of all pairs, with no need to be concerned about relative proportions of the initial investment.

A natural question that follows is how the capital is allocated for each pair. In theory, if the same value is being held in the long and short position of a pair, the capital earned from the short position can entirely cover the cost of entering the long position, and hence no initial financing is needed.[6] This is called a self-financing portfolio. Even though this assumption is reasonable in theory, in practice, it does not apply because a collateral is always required for borrowing the security being shorted, making a zero initial investment unfeasible. Therefore, the necessary investment corresponds to the collateral, which we assume to be the value of the

---

[6]Assuming the capital from the security being shorted is earned immediately after shorting.

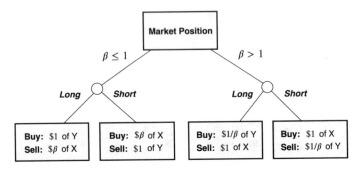

Fig. 5.8 Market position definition

security being shorted. This amount allows the investor to enter a long position with that same value. By financing long positions with proceeds from short positions, the investor is leveraging its investment. This leveraging approach is typical in hedge funds to increase the absolute return in the portfolio, and for that reason is considered in this work.

To simplify the calculations, it is particularly useful to consider a one dollar amount investment in each pair. This approach is undertaken by most authors in the field (Avellaneda and Lee [1], Caldeira and Moura [3], Dunis et al. [9], Gatev et al. [12], Rad et al. [19]), thus making it more suitable for comparison purposes as well. It is worth noticing that this approximation works on the assumption that an investor can buy fractions of trading ETFs, as the security itself is most probably not valued at exactly one dollar. Nonetheless, in this work, we adopted this assumption backed by the fact that the investor may always find the least common multiple between the two security prices that allow him to take an equal-sized position in both ETFs.

Contrarily to some research work that intends to be dollar-neutral, as Gatev et al. [12] and Dunis et al. [9], which invest $1 in the long position and $1 in the short position, the methodology followed in this work respects the cointegration ratio β between the two securities. Therefore, we propose that the value invested in X is β times the value invested in Y. In these conditions, to fixate a $1 initial investment, it needs to be assured that neither the long nor the short position costs more than $1. Formally, the condition is being imposed is that

$$\max(leg1, leg2) = \$1,$$

where $leg1$ and $leg2$ represent the capital invested in each leg of the pair. On this basis, we construct a framework illustrated in Fig. 5.8.

As the trading progresses, we consider that all the capital earned by a pair in the trading period is reinvested in the next trade. For instance, if the first time a pair trades it has a return of 5%, the second time a trade is opened for that pair, the initial capital will be $1.05, and not $1 as before. This mechanism facilitates the calculation of the final return.

Table 5.4 Transaction costs considered

	Commission costs	Market impact	Rental costs
Description	Value charged when buying or selling a security	Indicator that reflects the cost faced by a trader due to market slippage	Constant loan fee for the short position, payable over the life of each trade
Charge	8 bps	20 bps	1% per annum

5.5.2 Transaction Costs

We assure that all the results presented in this work account for transaction costs. The transaction costs considered are based on estimates from Do and Faff [7]. The authors perform an in-depth study on the impact of transaction costs in Pairs Trading. Conservative estimates of all the inherent costs are provided. These estimates are also adopted by other authors in the field, such as Huck and Afawubo [15]. The costs considered can be divided into three components, as described in Table 5.4. The charge is described relative to the position size.

It should be noted that these costs are estimated by Do and Faff [7] between 2009 and 2012, and their work indicates the trend is clearly favourable, as transaction costs tend to reduce in more recent years. Thus, if the results are robust to the considered transaction costs, at the time of writing they would only improve.

Commission and market impact costs must be adapted to account for both assets in the pair. The costs associated with one transaction are calculated as represented in Fig. 5.9.

5.5.3 Entry and Exit Points

One important subtlety regarding the trading execution is related to the prices considered when entering a position. In both the threshold-based and forecasting-based models, the system continuously monitors the prices to verify if a given threshold is triggered, in which case, a position is set. But since a trading system can not act instantaneously, there might be a small deviation in the entry price inherent to the delay of entering a position. It should be discussed how this situation is handled in the proposed simulation environment.

In line with Do and Faff [6], Dunis et al. [9], Gatev et al. [12], we assume a conservative delay of one period to enter a position. Hence, if a position is triggered at $t = i$, we only enter the position at time $t = i + 5$ min (recall the dataset considered has a frequency of 5 min). This way, we ensure the strategy is viable in practice. If it proves profitable under these conditions, then the results could only improve considering a shorter delay than five minutes, which is easily conceivable in practice.

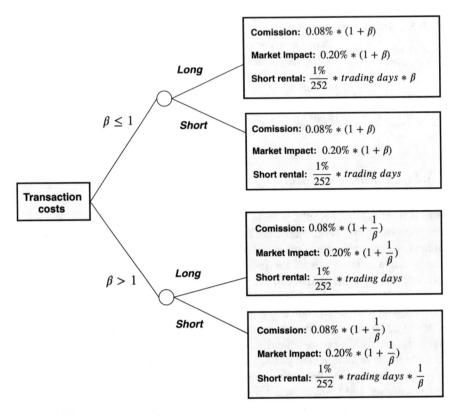

Fig. 5.9 Calculation of transaction costs

Concerning the definition of exit points, this work does not comprehend an implementation of a stop-loss system, under any circumstances. This means a position is only exited if one of two conditions is verified. Either the pair converges or the trading period ends. In the last scenario, a position that was recently entered might deteriorate the total return, even though it could turn out to be profitable in the future. Nevertheless, this is the best approximation of how a hedge fund would announce its profits at the end of a given period.

5.6 Evaluation Metrics

We proceed to describe the financial evaluation metrics deemed more appropriate to analyze the proposed strategies: the Return on Investment (ROI), the Sharpe Ratio (SR) and the Maximum Drawdown (MDD).

5.6.1 Return on Investment

By definition, the return on investment in its simplest form is given by the net profit divided by the initial investment, as follows:

$$ROI = \frac{\text{Net Profit}}{\text{Initial Investment}} \times 100. \tag{5.4}$$

Recall that the portfolio construction designed in this work establishes the initial required investment is always $1. This scenario has the advantage that the return at any point in time can also be interpreted as the net profit.

When calculating the returns of the entire portfolio, one question that occurs is whether the portfolio returns should be averaged over all the pairs selected for trading, or simply over the pairs that actually opened positions during the trading period. Gatev et al. [12] coined the former approach the return on committed capital (RCC), and the latter fully invested return (FII). This work considers the first approach. In spite of being more conservative, it accounts for the opportunity cost of hedge funds for having to commit capital to a strategy even if the strategy does not trade.

An additional subtlety should be noted. As Dunis et al. [9] claim, in a realistic scenario, the financial institution lending shares would pay an interest on the collateral. Therefore this interest should be summed to the strategy's net profit. However, we neglected this for the ease of calculation since it would represent only a very small portion of the net profit.

The returns, as presented, are calculated on a leveraged position, as explained in Sect. 5.5.1. In the unleveraged case, the initial capital corresponds to the initial gross exposure (value of the long position plus the value of the short position), and hence the returns would be slightly reduced.

5.6.2 Sharpe Ratio

Sharpe ratio is a risk-adjusted evaluation of return on investment first developed by Nobel laureate Sharpe [20]. It aims at measuring the excess return per unit of deviation in an investment, typically referred to as risk. The annual Sharpe ratio is calculated as

$$SR_{year} = \frac{R^{port} - R_f}{\sigma_{port}} \times \text{annualization factor.} \tag{5.5}$$

We proceed to describe each of the elements composing (5.5). First, R^{port} represents the expected daily portfolio returns. This may be calculated as the mean value of the portfolio returns, given by

$$R_t^{port} = \sum_{i=1}^{N} \omega_i R_t^i, \tag{5.6}$$

Table 5.5 Risk-free rates considered per test period

	Jan. 14–Dec. 14	Jan. 15–Dec. 15	Jan. 16–Dec. 16	Jan. 17–Dec. 17	Jan. 18–Dec. 18
R_f	0.03%	0.05%	0.32%	0.93%	1.94%

where R_t^i and ω_i represent the daily returns and the weight of the i-th pair in the portfolio of size N, respectively.[7] The risk-free rate, R_f, represents the expected rate of return of a hypothetical investment with no risk of financial loss. This is subtracted from returns earned by a given strategy to account for the interest that same cash amount could be generating instead, with no risk. In this work, we follow the common practice of setting the risk-free rate equal to the interest paid on the 3-month US government treasury bill.[8] Table 5.5 illustrates the risk-free annualized rates considered for each period tested. The values are obtained by averaging the 3-Month Treasury bill rate during the corresponding period. The data is collected from Board of Governors of the Federal Reserve System (US) [2].

The values presented in Table 5.5 must be converted to the expected daily returns, for coherence with (5.5). The term σ_{port} represents the volatility of the portfolio. This depends not only on the standard deviation of each security but also on the correlation among them, and is calculated as

$$\sigma_{\text{port}} = \sqrt{\sum_{i=1}^{N}\sum_{j=1}^{N} \omega_i \, \text{cov}(i,j)\omega_j} = \sum_{i=1}^{N}\sum_{j=1}^{N} \omega_i \sigma_{i,j} \omega_j. \tag{5.7}$$

Finally, the annualization factor enables expressing the estimated daily Sharpe ratio in annual terms. Expressing the annualized Sharpe ratio is convenient as this is a common practice and is consistent with the duration of the test periods. A common approximation to calculate the annual Sharpe ratio from the Sharpe ratio daily estimate consists of multiplying it by $\sqrt{252}$, where 252 is the number of trading days in a year. However, as pointed out by Lo [17], this is possible only under very special circumstances. Depending on the serial correlation of the portfolio returns, this approximation can yield Sharpe ratio values that are considerably smaller (in the case of positive serial correlation) or larger (in the case of negative serial correlation). The only case in which the approximation is truly valid is if the portfolio returns are independently and identically distributed (IID). This is a restrictive and often violated condition by financial data. Thus, for a more rigorous approximation, we adopt the correction factor proposed in Lo [17]. In this case, the portfolio returns serial correlation is measured, and a scale factor is applied in accordance. The scale factors are described in Appendix 2.

---

[7]For an equal weighted portfolio, $\omega_i = \frac{1}{N}$.

[8]The risk-free rate is a theoretical number since technically all investments carry some form of risk. Although the government can default on its securities, the probability of this happening is very low.

One disadvantage of SR is that the volatility measurement penalizes all volatility the same, regardless of if its upside (big positive returns) or downside volatility. Yet, large losses and large gains are not equally undesirable. A common alternative is the Sortino ratio, which is calculated just like the Sharpe ratio, with the exception that the volatility is calculated by taking only the standard deviation of the negative returns. Thus, the reader might be questioning why Sharpe ratio is adopted in this work instead of the Sortino ratio. The motivation is two-fold. First, Sharpe ratio is widely used in the Pairs Trading literature. It is the metric of choice in similar research work Avellaneda and Lee [1], Caldeira and Moura [3], Gatev et al. [12] which facilitates the comparison of the results. Secondly, the Sharpe ratio is independent of leverage. To justify this claim, we may quickly infer that leveraging up the committed capital n times would result in multiplying both the numerator and the denominator in Eq. (5.5) by n, meaning result remains unchanged.[9] This property is particularly convenient, given that the constructed portfolio is leveraged.

5.6.3 Maximum Drawdown

While the Sharpe ratio adjusts the returns for the risk taken, the Maximum Drawdown (MDD) indicates the maximum observed fall from a peak to a trough of a time series, before a new peak is attained. In this work, the Maximum Drawdown is calculated with respect to the account balance during the trading period. More formally, if we define $X(t)$, with $t \geq 0$ as representing the total account balance, the MDD is calculated as

$$\text{MDD}(T) = \max_{\tau \in (0,T)} \left[\max_{t \in (0,\tau)} \frac{X(t) - X(\tau)}{X(t)} \right]. \tag{5.8}$$

5.7 Implementation Environment

The code developed in this work is implemented from scratch using Python. The motivation for doing so is the vast amount of available resources that facilitate the implementation of the chosen algorithms, namely data mining procedures and data analysis. Furthermore, Python is, at the time of writing, the language of choice for Machine Learning related projects, which makes it more suitable from a collaboration point of view as well.

Some libraries are particularly useful in this work. First, *sci-kit learn* proves helpful in the implementation of PCA and the OPTICS algorithm. Second, *statsmodels* provides an already implemented version of the ADF test, useful for testing cointegration. Lastly, we make use of *Keras*, a Deep Learning library that provides the

---

[9]This approximation neglects the cost of funding leverage, which is considered a minor factor.

building blocks of a Neural Network. This way, we do not need to concern with the lower-level details, since most of the features described in Chap. 4 are already implemented.

Most of the simulation is run on a regular CPU (Intel Core i7 @ 3GHz). The exception is the training process of the Deep Learning models. These models involve a tremendous amount of matrix multiplications which result in long processing times when using the CPU. Fortunately, these operations can be massively sped up by taking advantage of the parallelization capabilities of a GPU. In this work, a GPU provided by Google Collaboratory (NVIDIA Tesla T4) is used to train the proposed architectures.

5.8 Conclusion

This chapter began with an overview of how the research experiment as a whole is conducted. We proceeded with a detailed description of the proposed dataset. We then explored the research environment suggested for each research stage in separate, describing the respective pairs selection conditions, trading setup, and test portfolios. Subsequently, we presented some important considerations from a simulation point of view. Namely, the way a portfolio is constructed, how transaction costs are accounted for, and the detail with which entry and exit points are defined. In addition, the most relevant metrics monitored in this work were also introduced. At last, we described some relevant aspects concerning the implementation environment.

Appendix 1

This appendix includes the total universe ETFs considered for trading in this work. For each ETF, the corresponding category and segment is presented (Tables 5.6, 5.7, 5.8 and 5.9).

Table 5.6 List of ETFs (1)

Ticker	Segment	Category	Ticker	Segment	Category
ACQ	Leveraged Commodities: Precious Metals Silver	Precious Metals	DBB	Commodities: Industrial Metals	Industrial Metals
AMJ	Equity: U.S. MLPs	Energy	DBC	Commodities: Broad Market	Broad Market
AMLP	Equity: U.S. MLPs	Energy	DBE	Commodities: Energy	Energy
AMU	Equity: U.S. MLPs	Energy	DBO	Commodities: Energy Crude Oil	Energy
AMZA	Equity: U.S. MLPs	Energy	DBP	Commodities: Precious Metals	Precious Metals
ATMP	Equity: U.S. MLPs	Energy	DBS	Commodities: Precious Metals Silver	Precious Metals
BCM	Commodities: Broad Market	Broad Market	DGAZ	Inverse Commodities: Energy Natural Gas	Energy
BNO	Commodities: Energy Crude Oil	Energy	DGL	Commodities: Precious Metals Gold	Precious Metals
BOIL	Leveraged Commodities: Energy Natural Gas	Energy	DGLD	Inverse Commodities: Precious Metals Gold	Precious Metals
CANE	Commodities: Agriculture Sugar	Agriculture	DGP	Leveraged Commodities: Precious Metals Gold	Precious Metals
CGW	Equity: Global Water	Agriculture	DGZ	Inverse Commodities: Precious Metals Gold	Precious Metals
CHIE	Equity: China Energy	Energy	DIG	Leveraged Equity: U.S. Energy	Energy
COPX	Equity: Global Metals & Mining	Industrial Metals	DJCI	Commodities: Broad Market	Broad Market
CORN	Commodities: Agriculture Corn	Agriculture	DJP	Commodities: Broad Market	Broad Market
CPER	Commodities: Industrial Metals Copper	Industrial Metals	DSLV	Inverse Commodities: Precious Metals Silver	Precious Metals
DBA	Commodities: Agriculture	Agriculture	DTO	Inverse Commodities: Energy Crude Oil	Energy

Table 5.7 List of ETFs (2)

Ticker	Segment	Category	Ticker	Segment	Category
DUG	Inverse Equity: U.S. Energy	Energy	FTGC	Commodities: Broad Market	Broad Market
DUST	Inverse Equity: Global Gold Miners	Precious Metals	FXN	Equity: U.S. Energy	Energy
DZZ	Inverse Commodities: Precious Metals Gold	Precious Metals	GASL	Leveraged Equity: U.S. Natural Gas	Energy
EMLP	Equity: U.S. MLPs	Energy	GASL	Leveraged Equity: U.S. Natural Gas	Energy
ENFR	Equity: U.S. MLPs	Energy	GASX	Inverse Equity: U.S. Natural Gas	Energy
DUST	Inverse Equity: Global Gold Miners	Precious Metals	GCC	Commodities: Broad Market	Broad Market
DZZ	Inverse Commodities: Precious Metals Gold	Precious Metals	GDX	Equity: Global Gold Miners	Precious Metals
EMLP	Equity: U.S. MLPs	Energy	GDXJ	Equity: Global Gold Miners	Precious Metals
ENFR	Equity: U.S. MLPs	Energy	GLD	Commodities: Precious Metals Gold	Precious Metals
ERX	Leveraged Equity: U.S. Energy	Energy	GLDI	Commodities: Precious Metals Gold	Precious Metals
ERY	Leveraged Equity: U.S. Energy	Energy	GLL	Inverse Commodities: Precious Metals Gold	Precious Metals
FCG	Equity: U.S. Natural Gas	Energy	GLTR	Commodities: Precious Metals	Precious Metals
FENY	Equity: U.S. Energy	Energy	GOEX	Equity: Global Gold Miners	Precious Metals
FILL	Equity: Global Oil & Gas Exploration & Production	Energy	GSG	Commodities: Broad Market	Broad Market
FIW	Equity: Global Water	Agriculture	GSP	Commodities: Broad Market	Broad Market
FRAK	Equity: Global Oil & Gas	Energy	IAU	Commodities: Precious Metals Gold	Precious Metals

Table 5.8 List of ETFs (3)

Ticker	Segment	Category	Ticker	Segment	Category
IEO	Equity: U.S. Oil & Gas Exploration & Production	Energy	MLPY	Equity: U.S. MLPs	Energy
IEZ	Equity: U.S. Oil & Gas Equipment & Services	Energy	NIB	Commodities: Agriculture Cocoa	Agriculture
IMLP	Equity: U.S. MLPs	Energy	NLR	Equity: Global Nuclear Energy	Energy
IXC	Equity: Global Energy	Energy	NUGT	Leveraged Equity: Global Gold Miners	Precious Metals
IYE	Equity: U.S. Energy	Energy	OIH	Equity: Global Oil & Gas Equipment & Services	Energy
JDST	Inverse Equity: Global Gold Miners	Precious Metals	OLEM	Commodities: Energy Crude Oil	Energy
JNUG	Leveraged Equity: Global Gold Miners	Precious Metals	OUNZ	Commodities: Precious Metals Gold	Precious Metals
KOL	Equity: Global Coal	Energy	PALL	Commodities: Precious Metals Palladium	Precious Metals
KOLD	Inverse Commodities: Energy Natural Gas	Energy	PDBC	Commodities: Broad Market	Broad Market
LIT	Equity: Global Metals & Mining	Industrial Metals	PHO	Equity: Global Water	Agriculture
MLPA	Equity: U.S. MLPs	Energy	PICK	Equity: Global Metals & Mining	Industrial Metals
MLPC	Equity: U.S. MLPs	Energy	PIO	Equity: Global Water	Agriculture
MLPG	Equity: U.S. MLPs	Energy	PPLT	Commodities: Precious Metals Platinum	Precious Metals
MLPI	Equity: U.S. MLPs	Energy	PSCE	Equity: U.S. Energy	Energy
MLPO	Equity: U.S. MLPs	Energy	PXE	Equity: U.S. Oil & Gas Exploration & Production	Energy
MLPX	Equity: U.S. MLPs	Energy	PXI	Equity: U.S. Energy	Energy

Table 5.9 List of ETFs (4)

Ticker	Segment	Category	Ticker	Segment	Category
PXJ	Equity: U.S. Oil & Gas Equipment & Services	Energy	UGA	Commodities: Energy Gasoline	Energy
REMX	Equity: Global Metals & Mining	Industrial Metals	UGAZ	Leveraged Commodities: Energy Natural Gas	Energy
RING	Equity: Global Gold Miners	Precious Metals	UGL	Leveraged Commodities: Precious Metals Gold	Precious Metals
RJA	Commodities: Agriculture	Agriculture	UGLD	Leveraged Commodities: Precious Metals Gold	Precious Metals
RJI	Commodities: Broad Market	Broad Market	UNL	Commodities: Energy Natural Gas	Energy
RJZ	Commodities: Broad Market Metals	Broad Market	URA	Equity: Global Nuclear Energy	Energy
RYE	Equity: U.S. Energy	Energy	USCI	Commodities: Broad Market	Broad Market
SCO	Inverse Commodities: Energy Crude Oil	Energy	USL	Commodities: Energy Crude Oil	Energy
SGDM	Equity: Global Gold Miners	Precious Metals	USLV	Leveraged Commodities: Precious Metals Silver	Precious Metals
SGOL	Commodities: Precious Metals Gold	Precious Metals	USO	Commodities: Energy Crude Oil	Energy
SIL	Equity: Global Silver Miners	Precious Metals	VDE	Equity: U.S. Energy	Energy
SILJ	Equity: Global Silver Miners	Precious Metals	WEAT	Commodities: Agriculture Wheat	Agriculture
SIVR	Commodities: Precious Metals Silver	Precious Metals	XES	Equity: U.S. Oil & Gas Equipment & Services	Energy
SLV	Commodities: Precious Metals Silver	Precious Metals	XLE	Equity: U.S. Energy	Energy
SLVO	Commodities: Precious Metals Silver	Precious Metals	XME	Equity: U.S. Metals & Mining	Industrial Metals

(continued)

Table 5.9 (continued)

Ticker	Segment	Category	Ticker	Segment	Category
SLVP	Equity: Global Silver Miners	Precious Metals	XOP	Equity: U.S. Oil & Gas Exploration & Production	Energy
SOYB	Commodities: Agriculture Soybeans	Agriculture	YMLI	Equity: U.S. MLPs	Energy
TAGS	Commodities: Agriculture	Agriculture	YMLP	Equity: U.S. MLPs	Energy
UCO	Leveraged Commodities: Energy Crude Oil	Energy	ZMLP	Equity: U.S. MLPs	Energy

Table 5.10 Scale factors for Sharpe ratios assuming returns follow an AR(1) process

ρ (%)	Aggregation value, q					
	12	24	36	48	125	250
40	2.36	3.27	3.98	4.58	7.35	10.37
30	2.61	3.65	4.44	5.12	8.23	11.62
20	2.88	4.04	4.93	5.68	9.14	12.92
10	3.16	4.45	5.44	6.28	10.12	14.31
0	**3.46**	**4.90**	**6.00**	**6.93**	**11.18**	**15.81**
−10	3.80	5.39	6.61	7.64	12.35	17.47
−20	4.17	5.95	7.31	8.45	13.67	19.35
−30	4.60	6.59	8.10	9.38	15.20	21.52
−40	5.09	7.34	9.05	10.48	17.01	24.11

Appendix 2

This appendix contains the table describing the scale factor correction method adopted in this work. The correction factor depends on the return's autocorrelation value ρ, and the aggregation value q, as described in Table 5.10. In this work, q is set to 250, since daily data corresponding to one year is being aggregated. The parameter ρ is measured for each portfolio's returns.

References

1. Avellaneda M, Lee JH (2010) Statistical arbitrage in the us equities market. Quant Financ 10(7):761–782
2. Board of Governors of the Federal Reserve System (US) (2019) 3-Month Treasury Bill: Secondary Market Rate. https://fred.stlouisfed.org/series/TB3MS, Accessed 2019-07-11
3. Caldeira J, Moura GV (2013) Selection of a portfolio of pairs based on cointegration: a statistical arbitrage strategy. Available at SSRN 2196391
4. Chan E (2013) Algorithmic trading: winning strategies and their rationale, vol 625. Wiley, New York
5. Chen H, Chen S, Chen Z, Li F (2017) Empirical investigation of an equity pairs trading strategy. Manag Sci
6. Do B, Faff R (2010) Does simple pairs trading still work? Financ Anal J 66(4):83–95. https://doi.org/10.2469/faj.v66.n4.1
7. Do B, Faff R (2012) Are pairs trading profits robust to trading costs? J Financ Res 35(2):261–287
8. Dunis CL, Laws J, Evans B (2006) Modelling and trading the gasoline crack spread: a non-linear story. Derivatives Use Trading Regul 12(1–2):126–145
9. Dunis CL, Giorgioni G, Laws J, Rudy J (2010) Statistical arbitrage and high-frequency data with an application to eurostoxx 50 equities. Liverpool Business School, Working paper
10. ETFcom (2019) ETFs - Tools, Ratings, News. https://www.etf.com/etfanalytics/etf-finder, Accessed 2019-06-30
11. Galenko A, Popova E, Popova I (2012) Trading in the presence of cointegration. J Altern Invest 15(1):85
12. Gatev E, Goetzmann WN, Rouwenhorst KG (2006) Pairs trading: performance of a relative-value arbitrage rule. Rev Financ Stud 19(3):797–827
13. Göncü A, Akyıldırım E (2016) Statistical arbitrage with pairs trading. Int Rev Financ 16(2):307–319
14. Huang CF, Hsu CJ, Chen CC, Chang BR, Li CA (2015) An intelligent model for pairs trading using genetic algorithms. Comput Intell Neurosci 2015:16
15. Huck N, Afawubo K (2015) Pairs trading and selection methods: is cointegration superior? Appl Econ 47(6):599–613. https://doi.org/10.1080/00036846.2014.975417
16. Krauss C, Do XA, Huck N (2017) Deep neural networks, gradient-boosted trees, random forests: statistical arbitrage on the S&P 500. Eur J Oper Res 259(2):689–702
17. Lo AW (2002) The statistics of sharpe ratios. Financ Anal J 58(4):36–52
18. Perlin M (2007) M of a kind: a multivariate approach at pairs trading
19. Rad H, Low RKY, Faff R (2016) The profitability of pairs trading strategies: distance, cointegration and copula methods. Quant Financ 16(10):1541–1558. https://doi.org/10.1080/14697688.2016.1164337
20. Sharpe WF (1994) The sharpe ratio. J Portf Manag 21(1):49–58

Chapter 6
Results

6.1 Data Cleaning

As a reminder, the initial set of securities considered in this work includes 208 different ETFs. Naturally, not all ETFs trade throughout every period considered. Furthermore, some violate the required conditions defined in Sect. 5.2.3.

   To illustrate the outcome of processing the original dataset, Table 6.1 describes the effect of each stage in the process. The first column represents the number of ETFs discarded due to missing values. This may be either because the ETF is not trading during the entire period or because there are sporadic missing values in the price series. The second column represents the number of ETFs, from the ones remaining, that do not verify the liquidity constraint imposed. The third column lists the number of ETFs containing outliers. On the left, the number of ETFs containing a single outlier in the price series is represented, in which case the value is corrected. On the right side, the number of ETFs for which the number of outliers is larger than one is detailed, in which case the ETF is discarded. Finally, the rightmost column indicates the percentage of initial ETFs analyzed to form pairs.

6.2 Pairs Selection Performance

This section focuses on analyzing the results addressing the first research question: "Can Unsupervised Learning find more promising pairs?". The periods under analysis correspond to those assigned to research stage 1 in Table 5.2.

6.2.1 Eligible Pairs

We start by presenting the number of pairs selected by the system under the three different clustering methodologies. These values are represented in Table 6.2.

© The Author(s), under exclusive license to Springer Nature Switzerland AG 2021
S. Moraes Sarmento and N. Horta, *A Machine Learning based Pairs Trading Investment Strategy*, SpringerBriefs in Computational Intelligence,
https://doi.org/10.1007/978-3-030-47251-1_6

Table 6.1 Data preprocessing results

Period	# of ETFs with insufficient data	# of ETFs with insufficient liquidity	# of ETFs with outliers (repaired–removed)	Percentage of total considered ETFs (%)
2009–2019	137	12	3–0	28
2012–2016	98	12	2–1	46
2013–2017	85	16	2–1	50
2014–2018	73	16	1–1	56
2015–2019	66	17	0–1	60

Table 6.2 Pairs selected for different clustering approaches

Formation period	2012–2015	2013–2016	2014–2017
No clustering			
Number of clusters	1	1	1
Possible combinations	4 465	5 460	6 670
Pairs selected	101	247	150
By category			
Number of clusters	5	5	5
Possible combinations	2 612	3 318	4 190
Pairs selected	59	51	51
OPTICS			
Number of clusters	9	13	12
Possible combinations	185	140	129
Pairs selected	39	40	18

As expected, the single cluster approach has a broader set of ETFs from which to select the pairs. Naturally, more pairs are selected using this setting. What is still left to asses is whether the pairs selected are all equally profitable.

We may observe that grouping the ETFs in five categories (according to the categories described in Sect. 5.2.2) results in a reduction in the number of possible pair combinations. This is not more evident due to the underlying unbalance across the categories considered. Since energy-linked ETFs represent nearly half of all ETFs, the combinations within this sector are still vast.

At last, the number of possible pair combinations when using OPTICS is remarkably lower. The results are obtained using five principal components to describe the reduced data, as it will be explained in Sect. 6.2.3.1. Also, the only OPTICS parameter to be defined, $minPts$, is set to three, since we empirically verify this value provides a reasonable distribution of clusters. Although the number of clusters is higher than when grouping by category, their smaller size results in fewer combinations. This does not just reduce the computational burden of performing multiple statistical tests but also reduces the probability of finding bogus pairs. The OPTICS reliance on the data is evidenced by the variation of the number of clusters across different periods.

6.2.2 Pairs Selection Rules

In this section, we intend to dig deeper into the details of the pairs selection procedure. The proposed selection process comprehends four criteria, as described in Sect. 3.5. There are essentially two aspects we would like to address in more detail. First, we want to examine what portion of the pairs passed the cointegration test but missed the spread Hurst exponent criterion. This result can be used as a proxy to estimate the real impact caused by the multiple comparisons problem and the influence of the Hurst exponent in handling this situation. Secondly, we are also interested in verifying the portion of identified mean-reverting pairs that do not comply with the two remaining imposed conditions, to gauge the importance of establishing those constraints. This is accomplished by illustrating the number of pairs filtered out by each condition. Focusing on the scenario in which no clustering is performed seems more appropriate as it results in more possible pair combinations and is, therefore, more representative.

Table 6.3 illustrates the pairs removed at each step of the selection procedure. It is important to note that the steps are performed in a sequential way. This means that each row represents the pairs eliminated from the subset resulting from the previous selection condition. Hence, it is expected that the majority of pairs are eliminated initially. Indeed, this is what we observe. Moreover, we can confirm the role of the Hurst exponent in eliminating pairs which passed the cointegration test, but their spread did not report a Hurst exponent lower than 0.5, thus not describing a mean-reverting process.

It can also be inferred from the results in Table 6.3 that some pairs are not eligible because their convergence duration is not compatible with the trading period, thus not verifying the half-life condition. Nevertheless, it is evident that the mean-crossing

Table 6.3 Pairs selection progress on an unrestricted pairs search environment

Formation period	2012–2015	2013–2016	2014–2017
Pairs eliminated per stage			
1. Cointegration	4 352	5 135	6 500
2. Hurst exponent	5	69	11
3. Half-life	7	9	9
4. Mean-crosses	0	0	0
Not eliminated	101	247	150
Total pairs	**4 465**	**5 460**	**6 670**

criterion is met for the entire subset of pairs that verified the previous conditions, which indicates that enforcing this last rule is redundant in this scenario.

In conclusion, we may say there is a visible influence of the Hurst exponent and the half-life constraints in filtering out pairs that would not have been discarded if simply cointegration had been enforced.

6.2.3 OPTICS Application Results

In this section, we intend to highlight some results from the application of the proposed technique to search for securities, based on the OPTICS algorithm. Namely, we present the outcome from varying the number of principal components used, and the composition of the clusters found. We also illustrate the application of DBSCAN in place of OPTICS, for comparison purposes.

6.2.3.1 Number of Principal Components

The first stage in the application of the proposed approach for clustering ETFs consists of defining an appropriate number of principal components to reduce the original data dimension. This procedure must achieve a balance between two desirable but incompatible features. On the one hand, more accurate representations may be obtained when the number of principal components increases, since the transformed data explains more of the original variance. On the other hand, increasing the number of components brings along more components expressing random fluctuation of prices, which should be neglected. In addition, it also contributes to the curse of dimensionality, as described in Sect. 3.3.

Table 6.4 illustrates how the clusters and the pairs found vary as the number of principal components changes. The numbers described are obtained in the period ranging from January 2014 to December 2018. This period is chosen for illustration arbitrarily since the behaviour is found to be consistent among the three periods used in this research stage. The last row emphasizes the pairs which contain constituents from distinct categories, thus emphasizing part of the diversity provided by this method. It is clear from the results disclosed in Table 6.4 that the impact of varying the number of principal components within the experimented interval is minimal. The demonstrated consistency regarding the number of clusters and their average size for a different number of principal components is a direct consequence of the robustness provided by the OPTICS algorithm. If the number of principal components is selected within an acceptable range, the OPTICS algorithm adapts its parameters automatically to fit best the data described by different dimensions. Obtaining the same result using DBSCAN is considerably harder, as the investor would have to find the best value for the parameter ε in each scenario.

In the previous paragraph, we described how the clustering varies within a range of acceptable dimensions. As it was motivated at the end of Sect. 3.3, we considered as

Table 6.4 Clustering statistics when varying number of principal components

Number of principal components	5	8	12	15
Number of clusters	12	9	10	9
Average cluster size	5.2	5.3	3.4	5.0
Pairs selected	18	16	4	16
Pairs selected from distinct categories	5	6	0	5

Table 6.5 Clustering statistics using large number of principal components

Number of principal components	50	75	100
Number of clusters	3	3	2
Average cluster size	5.6	4.3	3
Pairs selected	6	3	0
Pairs selected from distinct categories	3	0	0

acceptable a number of features up to 15, to protect from the curse of dimensionality. We find relevant to conduct an exploratory analysis concerning how the algorithm performance is affected for considerably higher dimensions. For this purpose, we simulate the same scheme but now using more principal components. The results are presented in Table 6.5. Interestingly, the algorithm is still capable of finding pairs when using 50 and even 75 dimensions, but naturally with fewer clusters formed. However, for 100 dimensions, the algorithm is incapable of detecting any pair, as one would expect to witness eventually. It should be emphasized that we do not proceed to confirm the reliability of the identified pairs from a trading perspective in these conditions.

To conclude, we settled the number of principal components to 5. As demonstrated above, this parameter should not affect distinctively the results obtained as long as it is within an acceptable dimension range. We adopt 5 dimensions since we find adequate to settle the ETFs' representation in a lower dimension provided that there is no evidence favouring higher dimensions.

6.2.3.2 Clusters Composition

In this section, we intend to validate the clusters formed by the OPTICS algorithm and get an insight into their composition. Illustrating the process for all the periods considered would be very extensive. Because the process is common across all the periods, we decide to proceed with the analysis of the period of January 2014 to December 2017 in this section and include the information for the remaining periods in Appendix 1.

To illustrate the clustering results it would be particularly appealing to show a representation of the generated clusters in the two-dimensional space. However, the data is described by five dimensions.[1] To tackle this issue, we propose the application of the T-distributed Stochastic Neighbor Embedding algorithm (t-SNE) Maaten and Hinton [6]. Briefly, t-SNE can be described as a nonlinear dimensionality reduction technique well-suited for embedding high-dimensional data for visualization in a low-dimensional space of two (or three) dimensions. The algorithm models each high-dimensional object using two (or three) dimensions in such a way that similar objects are modelled by nearby points and dissimilar objects by distant points with high probability.[2]

Using the technique just described, a two-dimensional map can be attained. Figure 6.1 illustrates the clusters formed during the formation period, in the transformed space. Each cluster is represented by a different colour. The ETFs not grouped are represented by the smaller grey circles, which were not labelled to facilitate the visualization. It should be noted that Fig. 6.1 is not displaying the entire set of ETFs, as two of them are placed in distant positions, and their representation makes it harder to visualize the clusters with the desired level of granularity. The existence of such outliers, from a data distribution point of view, confirms the importance of using a clustering technique robust to such cases.

Although the clusters formed look plausible, based solely on the current evidence, it may be premature to commentate on their integrity. To get a better insight of the securities composing each cluster, we suggest analyzing the corresponding price series. For this purpose, we select four clusters for visualization. Figure 6.2 illustrates the logarithm of the price series of each ETF. The price series illustrated result from subtracting the mean of the original price series, to facilitate the visualization.

Starting with Fig. 6.2a, it is interesting to observe how close the series follow each other, making it hard to differentiate them. One question that emerges is whether there is a fundamental reason that justifies this behaviour. There is. The four securities identified do not just belong to the same category, Precious Metals, but are also linked to Gold. This is the first evidence that the OPTICS-based approach can go one level deeper in grouping ETFs, as it is capable of detecting not just categories, but specific segments. A similar situation is encountered in Fig. 6.2b. In this case, the

---

[1] Recall we selected five principal components when applying PCA.

[2] For more information concerning the t-SNE algorithm the interested reader may refer to the original paper Maaten and Hinton [6].

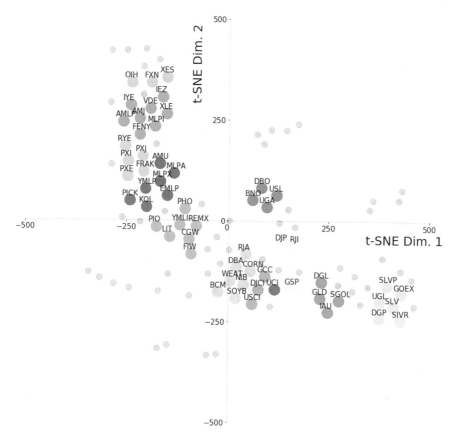

Fig. 6.1 Application of t-SNE to the clusters generated by OPTICS during Jan 2014 to Dec 2016

securities identified also belong to the same segment, the Equity US: MLPs, which is a component of the entire Energy sector.

Figure 6.2c, d demonstrate the OPTICS clustering capabilities extend beyond selecting ETFs within the same segment. In cluster 3, we may find ETFs belonging to the Agriculture sector (CGW, FIW, PHO, and PIO), ETFs associated with Industrial Metals (LIT and REMX) and ETFs in the Energy sector (YMLI). There is a visible relation among the identified price series, even though they do not all belong to the same category. The same applies to cluster 4, which is composed of three different categories, Energy (KOL), Industrial Metals (PICK) and Broad Market (UCI).

To conclude, we confirm the clusters formed accomplished its intent of combining the desired traits from the two other techniques studied. Namely, a tendency to group subsets of ETFs from the same category while not impeding clusters containing ETFs from different categories.

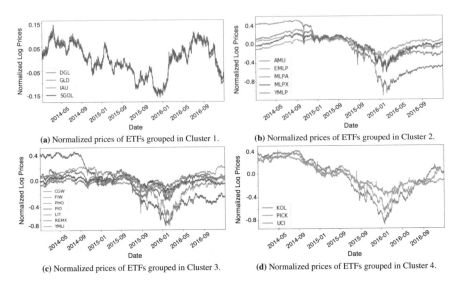

(a) Normalized prices of ETFs grouped in Cluster 1.

(b) Normalized prices of ETFs grouped in Cluster 2.

(c) Normalized prices of ETFs grouped in Cluster 3.

(d) Normalized prices of ETFs grouped in Cluster 4.

Fig. 6.2 Price series composition of some clusters formed in the period from Jan 2014 to Dec 2016

6.2.3.3 Comparison with DBSCAN

As attentively motivated at the end of Sect. 3.4.3, the OPTICS algorithm is particularly advantageous in comparison with DBSCAN for two reasons. It handles well clusters of varying density within the same dataset, and it avoids the cumbersome process the investor would have to go through to find the appropriate value for ε. To illustrate how this is evident in practice, we proceed to represent the clusters that would have been formed using DBSCAN, during the same formation period, for a varying parameterization of ε in Fig. 6.3.

From the observation of the three scenarios in Fig. 6.3, the first aspect that stands out is the high sensibility of the algorithm to variations in the parameter ε. This aspect reinforces that the investor must be very cautious when choosing the parameteriza-

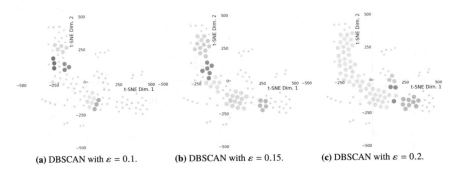

(a) DBSCAN with $\varepsilon = 0.1$.   **(b)** DBSCAN with $\varepsilon = 0.15$.   **(c)** DBSCAN with $\varepsilon = 0.2$.

Fig. 6.3 Applying DBSCAN with varying ε to the formation period from Jan 2014 to Dec 2016

tion. Furthermore, from the three configurations presented, how should the investor select ε ? It might not be straightforward to decide which combination yields the most promising results and leaving this decision to the investor might be too demanding. The OPTICS algorithm, however, allows the automation of this process.

6.2.4 Trading Performance

In this section, we analyze how the selected pairs under the three distinct clustering conditions perform from a trading point of view. This enables validating which clustering technique generates the most promising portfolio.

We start by revealing the results obtained with the threshold-based trading model, as proposed in Sect. 5.3.2, during the validation periods included in Table 6.6. The results presented in this table should be analyzed with caution because there is an implicit bias arising from the fact that the strategy is being examined in the same period used to identify the pairs being tested, which does not resemble a real trading scenario. Hence, the results are not truly representative of the real performance. If on the one hand, one might expect the results to be more satisfactory, due to the extra confidence on the robustness of the pairs being used, on the other hand, they may decline, since we cannot eliminate pairs that have not been profitable in the past, as there are no available records yet. Nevertheless, the validation records are relevant to support the construction of the test portfolios, which use these results as a heuristic for selecting pairs. For this reason, we decide to present them. One key idea to retain from Table 6.6 is that the strategy is profitable in every scenario, with a satisfactory margin. Therefore we can confirm the success of the pairs selection process in selecting profitable pairs up to this point.

Table 6.7 unveils the results concerning the unseen data, the test periods. In this case, for each test period, we describe the results when adopting any of the three implemented test portfolios, described in Sect. 5.3.3. As a quick reminder, the portfolio identified by number 1 comprehends the entire set of pairs identified in the formation period, portfolio number 2 contains only the pairs that had a satisfactory performance in the validation period, and portfolio 3 comprises the ten pairs which achieved the best performance in the validation period. To aggregate the information in a more concise way, the average over all years and portfolios is described on the rightmost column. Note also that the three evaluation metrics described in Sect. 5.6 are accentuated, to differentiate from the remaining, less critical, descriptive statistics.

We can confirm the persisting profitability in any of the environments tested. This result further corroborates the idea that the pairs selection procedure is robust. Besides, we may assert that if an investor is focused on obtaining the highest possible ROI, regardless of the incurred risk, performing no clustering is particularly appealing. Except for 2015, when the OPTICS based strategy performed slightly better in this sense, the no clustering approach obtains higher ROI overall. However, if the investor is also concerned with the risk its portfolio is exposed to, the remainder

Table 6.6 Validation results for each pairs grouping technique

Validation period	2014	2015	2016
No clustering			
SR	4.22	5.17	5.66
ROI (%)	11.3	19.8	19.9
MDD (%)	1.35	2.37	1.24
Profitable pairs (%)	76	90	95
# trades (positive-negative)	276 (198-78)	581 (511-70)	384 (324-60)
Category			
SR	2.07	3.81	5.24
ROI (%)	6.94	11.45	27.35
MDD (%)	1.71	1.80	2.46
Profitable pairs (%)	68	86	92
# trades (positive-negative)	167 (104-63)	136 (112-24)	145 (127-18)
OPTICS			
SR	4.09	5.70	3.97
ROI (%)	8.33	10.96	12.65
MDD (%)	0.97	0.55	1.39
Profitable pairs (%)	87	88	89
# trades (positive-negative)	105 (85-20)	157 (131-26)	57 (45-12)

indicators suggest that clustering using OPTICS may achieve a better trade-off. We proceed to analyze why.

The OPTICS-based approach is capable of generating the highest average portfolio Sharpe ratio of 3.79, in comparison with 3.58 obtained when performing no clustering or 2.59 when grouping by category. It also shows more consistency with respect to the portion of profitable pairs in the portfolio, with an average of 86% profitable pairs, against 80% when grouping by category and 79% when performing no clustering at all. It is interesting to observe that the only circumstances under which 100% of the pairs in the portfolio turned out to be profitable correspond to the situation that makes use of the OPTICS algorithm. This happened during two different periods, 2015 and 2017, when using portfolio 3. This stresses the consistency provided by this clustering approach.

Concerning the maximum drawdowns, we notice that this metric evidences some instability when grouping by category or not grouping at all. The results indicate that when these strategies use the ten best-performing pairs from the validation period (portfolio 3), the portfolio maximum drawdown amplitude tends to increase markedly. This is especially evident in 2017, in which case the MDD has nearly the same amplitude as the ROI. Contrarily, the OPTICS based strategy proves capable

Table 6.7 Test results for each pairs grouping technique

Test period	2015			2016			2017			AVG.
Test portfolio	1	2	3	1	2	3	1	2	3	–
No clustering										
SR	3.53	4.12	3.32	3.96	4.51	3.56	4.08	4.05	1.11	3.58
ROI (%)	10.4	12.4	17.4	24.8	26.3	26.0	11.9	12.4	11.5	17.0
MDD (%)	1.42	0.97	2.59	2.05	1.98	2.65	1.33	1.38	9.28	2.63
Total pairs	101	77	10	247	223	10	150	141	10	108
Profitable pairs (%)	70	80	70	86	86	90	69	70	90	79
Total trades	229	173	17	411	361	15	212	195	14	181
Profitable trades	180	147	15	369	329	15	172	162	12	156
Unprofitable trades	49	26	2	42	32	0	40	33	2	25
Category										
SR	1.56	2.39	3.75	3.48	3.82	3.09	2.17	2.14	0.89	2.59
ROI (%)	5.52	9.38	17.8	13.6	13.9	20.4	7.86	8.42	8.31	11.7
MDD (%)	1.77	1.82	2.09	2.06	2.26	4.56	2.47	2.67	8.91	3.18
Total pairs	59	40	10	51	44	10	51	47	10	36
Profitable pairs (%)	64	85	90	86	86	90	65	64	90	80
Total trades	154	108	39	107	83	20	64	54	13	71
Profitable trades	112	89	36	92	73	19	49	43	12	58
Unprofitable trades	42	19	3	15	10	1	15	11	1	13
OPTICS										
SR	4.05	3.84	5.08	4.72	4.79	3.80	2.75	2.83	2.27	3.79
ROI (%)	12.5	13.5	23.5	10.5	11.9	15.2	7.36	8.38	9.98	12.5
MDD (%)	1.37	1.66	1.30	0.80	0.83	1.46	1.21	1.35	2.35	1.37
Total pairs	39	34	10	40	35	10	18	16	10	24
Profitable pairs (%)	82	82	100	80	83	90	78	81	100	86
Total trades	161	147	68	87	78	30	24	22	17	70
Profitable trades	140	128	67	72	66	27	21	20	17	62
Unprofitable trades	21	19	1	15	12	3	3	2	0	8

of maintaining a favourable MDD in every period, with no irregularities. This is confirmed by the lower average MDD, which emphasizes the robustness provided by this technique.

Lastly, we can verify that since the OPTICS algorithm only tends to group pairs that seem more sound, the system is less likely to identify pairs that achieve a very good validation performance and see it deteriorate in the test period. This aspect may be validated from the analysis of the Sharpe ratio progression from the validation to the test results. While on the validation set (Table 6.6) the OPTICS based approach is only able to outperform the other strategies in 2015, in the test set it surpasses the remaining techniques most of the times.

Curiously, regarding the two standard grouping techniques, we observe that grouping the clusters by category yields poorer results than those obtained when not grouping at all. This suggests that, in the absence of OPTICS, the investor might be better off not performing any grouping, instead of limiting pairs to be formed by securities from the same sector.

On a different note, the results indicate that there is no indisputable best way of constructing the test portfolio. Evaluating by the Sharpe ratio, the portfolio with the best performance depends on the period and the clustering technique being used. As every portfolio outperforms the other two at least once, the results are inconclusive. However, evaluating in terms of ROI, we may infer that using information from the validation performance for constructing the test portfolio is a useful heuristic after all, as portfolio 2 and 3 outperform portfolio 1 in every case.

6.3   Forecasting-Based Trading Model Performance

This section focus on analyzing the results that let us answer the second research question: "Can a forecasting-based trading model achieve a more robust performance?". To accomplish this, we evaluate how the forecasting-based model, introduced in Chap. 4, performs when compared with the threshold-based model, simulated so far. In doing so, we gradually describe how the simulation progresses, from the pairs identification until the trading itself.

6.3.1   Eligible Pairs

We proceed to describe the selected pairs during this research stage. In this case, only the OPTICS clustering is applied to cluster the candidate pairs. This approach is chosen due to its already demonstrated ability to form robust portfolios. The clustering is then followed by the regular pairs selection procedure, just as before. Table 6.8 presents some relevant statistics regarding the pairs found for both periods used in this second research stage.

Only 5 pairs are deemed eligible for trading based on the formation period from Jan 2009 to Dec 2017, considerably fewer than those found in the other periods analyzed. This is not surprising since there are just 58 active ETFs throughout this period, as detailed in Table 5.2. In addition, the portion of cointegrated ETFs during

Table 6.8 Pairs selected during research stage 2, using OPTICS

Formation period	2009–2017	2015–2017
Number of clusters	5	17
Possible combinations	48	129
Pairs selected	5	19

such a long period must be smaller. Nonetheless, this does not represent a threat since selecting fewer pairs can be desirable in this setting. Because training the Deep Learning forecasting models is computationally very expensive, the computation time can be greatly reduced by considering just a small number of pairs. This also provides a good opportunity to take a closer look at the identified ETFs, one by one. Table 6.9 lists the selected pairs, along with the corresponding segment, and their description. Both the segment and the description information are retrieved from ETF.com [3].

It is interesting to notice that two out of five pairs belong to different categories. To get a better insight into the underlying motive, we analyze the DBA-GCC pair in more detail. As GCC is a Broad Market type of ETF, meaning that it follows the commodity market as a whole, we gather more information from ETF.com [4] regarding its sector decomposition, described in Table 6.10.

By analyzing Table 6.10, one can observe that the major component driving GCC is Agriculture. Thus, an equilibrium between GCC and DBA can be interpreted as a consequence of major exposure to the same sector. The deviations from the equilibrium can be potentially justified by market changes in the remaining sectors.

Revisiting Table 6.8, now focusing on the period of January 2015 to December 2018, we find 19 pairs eligible for trading. This number is consistent with the values found in the previous periods of the same duration. These pairs are used to gauge the performance of the benchmark trading model, applied in its standard interval range (4 years, as applied in Sect. 6.2).

6.3.2 Forecasting Algorithms Training

The forecasting algorithms are trained to fit the spreads from the five pairs selected during the formation period of January 2009 to December 2017, represented in Fig. 6.4. It is interesting to observe that the spreads look stationary, as expected. Moreover, there is a noticeable difference in the volatility among the time series, which further supports the idea that data-driven model parameters, as the proposed thresholds for the forecasting-based trading model, seem appropriate in these conditions.

Having described the training data, we advance with the forecasting architectures experimented. The architectures are made more complex until there is no more

Table 6.9 Candidate pairs identified from Jan 2009 to Dec 2017

Pairs	Segment	Description
PXE	Equity: U.S. Oil & Gas Exploration & Production	PXE tracks an index of U.S. energy exploration and production companies selected and weighted by growth and value metrics
PXI	Equity: U.S. Energy	PXI tracks an index of U.S. energy firms selected and weighted by price momentum
DBA	Commodities: Agriculture	DBA tracks an index of 10 agricultural commodity futures contracts. It selects contracts based on the shape of the futures curve to minimize contango
GCC	Commodities: Broad Market	GCC tracks an equal-weighted index of 17 commodities. It uses futures contracts averaged across the nearest 6 months of the futures curve to maintain its exposure and rebalances daily
GCC	*same as previous row	*same as previous row
RJA	Commodities: Agriculture	RJA tracks a consumption-based index of agricultural commodities chosen by the RICI committee. It gains exposure using front-month futures contracts
GSP	Commodities: Broad Market	GSP tracks a broad index of commodities futures contracts weighted according to global production
RJI	Commodities: Broad Market	RJI tracks a consumption-based index of commodities chosen by the RICI committee. It gains exposure using front-month futures contracts
DGP	Leveraged Commodities: Precious Metals Gold	DGP tracks the performance of futures contracts relating to gold
UGL	Leveraged Commodities: Precious Metals Gold.	UGL provides 2x exposure to the daily performance of gold bullion as measured by the fixing price, in US dollars

Table 6.10 GCC sector breakdown

Sector	GCC portfolio breakdown (%)
Agriculture	57.15
Precious metals	21.44
Livestock	14.29
Industrial metals	7.12
Energy	0.00

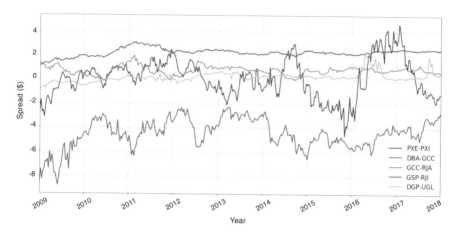

Fig. 6.4 Spread from the five pairs identified between Jan 2009 and Dec 2018

evidence of improvement. A total of 31 forecasting model architectures are implemented, meaning a total of 155 models are trained (31 architectures × 5 spreads).
From the 31 implemented architectures, 9 correspond to the configurations experimented with the ARMA model, 8 with the LSTM based model and 14 with the LSTM
Encoder-Decoder. The implemented configurations are described in Appendix 3. The
cost function that each forecasting algorithm aims to optimize is the MSE. To see the
details regarding MSE as well as other forecast measures used in this work, please
refer to Appendix 2.

It is worth adding that although both the parametric and the non-parametric
approaches use the formation period for training, the methodology slightly differs for
each case. While the ARMA model uses the entire formation period to learn the combination of coefficients that minimizes the forecasting error, the LSTM, and LSTM
Encoder-Decoder use only the training period for that purpose. In this last case, the
validation set is considered separately to control overfitting. Through early-stopping,
as described in Sect. 4.7.3, the system continuously monitors the validation score and
stops the training process if signs of overfitting are detected. Figure 6.5 illustrates one
typical training scenario with an example from the LSTM Encoder-Decoder fitting
the spread resulting from the pair DGP-UGL.

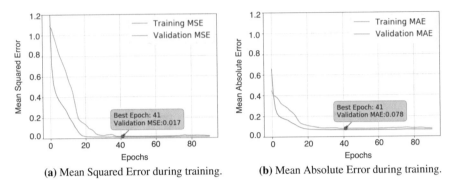

(a) Mean Squared Error during training.   (b) Mean Absolute Error during training.

Fig. 6.5 Encoder-Decoder training loss history for DGP-UGL spread

On Fig. 6.5a, we may observe how the training and validation MSE evolve. As expected, they both decrease as the training progresses. But the algorithm seems to reach its optimal configuration at epoch 41. From that point on, there is no improvement in the validation MSE, despite the still decreasing training MSE. This suggests the algorithm is overfitting. The figure represents a total of 91 epochs (although not visible) because the early-stopping is configured with a patience of 50 epochs. Thus, the training continues for 50 epochs and stops at the 91st, since no better score is obtained in the meanwhile. It is also worth observing the image on the right, Fig. 6.5b, which depicts the evolution of the MAE. Although the model is configured to minimize the MSE, it is visible that the MAE progression displays a similar pattern.

6.3.3 Forecasting Performance

Having described the selected pairs and how the training is conducted for each forecasting approach, we may proceed with the analysis of the forecasting results obtained. An extensive list of results concerning the forecasting score of all the implemented architectures is presented in Appendix 3. In this section, we focus on comparing the results from the best performing architectures.

It is important to clarify that we consider the best architectures as those which minimize the validation mean squared error. The reason for using the validation score as a reference is that it resembles the information available to the investor prior to start trading, the point at which this decision must be taken. The mean squared error is the targeted metric used for comparison purposes as this is the loss function being minimized by the models. According to these criteria, the best performing architectures are presented in Table 6.11. Also, the results described for each model correspond to the average forecasting performance across the five spreads' forecasts.

Table 6.11 Forecasting results comparison

Model	Naïve		ARMA	LSTM	LSTM Encoder-Decoder	
Configuration	–		p: 8	input length: 24	input length: 24	
	–		q: 3	hidden layers: 1	encoder nodes: 15	
	–		–	hidden nodes: 50	decoder nodes: 15	
Time step	(t+1)	(t+2)	(t+1)	(t+1)	(t+1)	(t+2)
Parameters	–		11	10 651	3 016	
Validation						
MSE (E-03)	1.87	3.34	1.51	1.69	1.71	2.05
RMSE (E-02)	3.69	4.94	3.00	3.28	3.32	3.60
MAE (E-02)	1.50	2.24	1.78	2.04	2.13	2.45
Test						
MSE (E-03)	2.60	4.47	2.26	3.35	4.31	9.03
RMSE (E-02)	3.89	5.14	3.34	4.30	5.21	7.50
MAE (E-02)	1.68	2.61	1.96	3.08	3.94	5.91

Inspecting the results from Table 6.11, we may validate that all the implemented models are capable of outperforming the naive implementation (defined in Sect. 5.4.1) during the validation period. Curiously, we note that the LSTM-based models do not manage to surpass the ARMA model, at least with respect to the chosen metrics. Also, the results obtained in the test set show signs of overfitting, besides the efforts taken in that regard, as the LSTM-based models are no longer superior to the naive model.

These results naturally make us question the suitability of LSTMs under this work's conditions, which leads us to dig deeper into the literature. Although the fitness of different forecasting algorithms strongly depends on the dataset, the literature evidence that there is no consensus concerning the suitability of LSTMs for time series forecasting.

On the one hand, some research work supports the dominance of RNNs over simpler techniques such as multi-layer perceptrons (MLP) or auto-regressive models. For instance, Olivier et al. [7], Tenti [9] allege these networks are generally better than MLP networks, although they do suffer from long computational times. Also, Siami-Namini et al. [8] investigates whether LSTMs are superior to the traditional algorithms, such as the ARIMA model, and the empirical studies conducted confirm the hypothesis. These type of considerations motivated the application of LSTM-based models in this work in the first place.

On the other hand, some authors as Gers et al. [5] claim that common time series problems found in the literature are often conceptually simpler than many tasks already solved by LSTMs, and that more often than not, they do not require RNNs at all, because a few recent events convey all relevant information about the next event. More specifically, the authors find evidence that the time window based MLP outperforms the LSTM approach on certain time series prediction benchmarks solvable by looking at a few recent inputs only, thus concluding that LSTM's special capabilities are not necessary for such circumstances. Our results are aligned with this point of view. Although an MLP model is not implemented, the results clearly indicate that under this problem conditions, the traditional ARMA model proves more robust with respect to the metrics identified in Table 6.11. This suggests the extra complexity provided by the LSTM-based models might be unnecessary. In the next section, we intend to analyze how this impacts trading profitability.

6.3.4   Trading Performance

We start by recalling that the primary motivation for pursuing the forecasting-based pairs trading approach is to make the portfolio more robust, by reducing the number of days the portfolio sees its value decrease. To evaluate the effectiveness of the proposed methodology, we proceed to describe the results obtained by the forecasting-based trading model using the different forecasting algorithms, during the period of January 2009 to December 2018. The results, in Table 6.12, are set side by side with the results obtained using the standard threshold-based model. It is important to remind once more that the results in the validation set are presented for completeness. Its real duty is to extract the pairs that prove capable of generating profits in the validation period, rather than testing the strategy itself.

Focusing on the test results, which resemble a practical trading environment, we can validate that the period of portfolio decline is considerably shorter when using the forecasting-based models, in comparison with the standard trading model. The longer period corresponds to the LSTM Encoder-Decoder, of 21 days, that is still less than 25% of the standard trading model's total decline period, of 87 days. However, that is not all there is to it, as this comes at the expense of a worse performance in terms of total return on investment and portfolio Sharpe ratio.

We suspect the long required formation period is also responsible for the visible profitability decline. Therefore we proceed to analyze the standard trading model in a regular 3-year-long period, this time focused on the profitability indicators, to have a baseline for comparison. The results are displayed in Table 6.13.

By comparison, the performance in the 10-year-long period seems greatly affected by the long required duration, suggesting the less satisfactory returns emerge not merely as a consequence of the trading model itself, but also due to the underlying time settings. Following this line of reasoning, if the forecasting-based model performance increases in the same proportion as the standard trading model when reducing the formation period, the results obtained could be much more satisfactory. The formation period may be reduced, for example, by using higher frequency data, thus maintaining the same level of data points.

Table 6.12 Trading results comparison using a 9-year-long formation period

Trading model	Standard	ARMA based model	LSTM based model	LSTM Encoder Decoder based model
Parameters	see Table 5.3	$\alpha_S = Q_{f^-(0.20)}$ $\alpha_L = Q_{f^+(0.80)}$	$\alpha_S = Q_{f^-(0.10)}$ $\alpha_L = Q_{f^+(0.90)}$	$\alpha_S = Q_{f^-(0.10)}$ $\alpha_L = Q_{f^+(0.90)}$
Validation				
SR	1.76	0.58	0.65	0.30
ROI (%)	5.09	2.21	1.74	1.40
MDD (%)	1.54	2.40	0.82	1.62
Days of decline	85	78	51	70
Trades (Pos.-Neg.)	17 (14-13)	345 (135-210)	130 (60-70)	187 (80-107)
Profitable pairs	4	3	3	3
Test				
SR	1.85	1.22	0.50	0.98
ROI (%)	6.27	5.57	2.93	4.17
MDD (%)	1.43	0.73	0.47	1.19
Days of decline	**87**	**11**	**2**	**21**
Trades (Pos.-Neg.)	149 (89-60)	34 (22-12)	8 (6-2)	17 (14-3)
Profitable pairs	3	3	2	2

Table 6.13 Trading results for standard trading model using a 3-year-long formation period

Trading model	Standard
Parameters	see Table 5.3
Validation	
SR	3.76
ROI (%)	14.1
MDD (%)	1.36
Days of decline	100
Identified pairs	19
Trades (Pos.-Neg.)	45 (43-2)
Profitable pairs	18
Test	
SR	**3.41**
ROI (%)	**11.3**
MDD (%)	1.12
Days of decline	89
Trades (Pos.-Neg.)	30 (26-4)
Profitable pairs	13

On a different note, as we suspected from the forecasting results presented in Sect. 6.3.3, the LSTM-based models do not manage to surpass the ARMA-based trading model. Yet, the fact that the LSTM Encoder-Decoder clearly exceeds the simpler LSTM model indicates there is some potential to using multi-step forecasting in this setting. Thus, it is left for future research to evaluate how an adapted multi-step prediction using ARMA would have performed.

6.3.5   Implementation Remarks

From a practical point of view, it is relevant to describe the training duration associated with fitting the forecasting algorithms. Naturally, the duration increases with the complexity of the architectures experimented. In our analysis, we focus on the models which obtained the best forecasting score. Table 6.14 introduces the training duration for two different test environments. In the first one, a regular CPU is used. In the second case, the K80 GPU provided by Google Collaboratory is exploited. In this case, the standard LSTM layers are replaced with a GPU-accelerated version, the *cuDNNLSTM* layers.

The values presented in Table 6.14 correspond to the total duration of fitting the algorithms for five times (for the five pairs). The ARMA model requires only the CPU since the training period is relatively short. As for the LSTM-based models, with the number of epochs fixed to 500 and using a batch of size 512, the GPU provides valuable gains.

6.3.6   Alternative Methods

In Sect. 4.2, two different methodologies to define more precise entry points were proposed. We ran some experiments until we had enough evidence to dismiss them. In this section, we intend to shed some light on why these approaches might not be so adequate in this scenario.

One of the alternative approaches described consists in attempting to predict the return series and set a position accordingly. This approach has been applied in the literature with relative success, which makes it a promising candidate Dunis et al. [1, 2]. However, the results could not be replicated in this work's setting. When trying to

Table 6.14 Forecasting algorithms training duration

Environment	CPU	GPU
Forecasting algorithm		
ARMA	00h 33m	–
LSTM	19h 25m	02h 47m
LSTM Encoder-Decoder	09h 43m	02h 05m

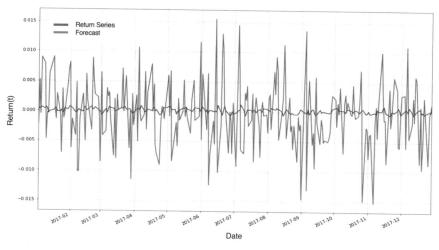

Fig. 6.6 Return series forecast

forecast the return series, we verify that the signal is extremely hard to predict, given its constant change of direction. Because the signal is highly volatile, and it constantly crosses its mean, the neural network tends to be better off simply predicting the mean value, apart from some small deviations, as represented in Fig. 6.6.

This phenomenon makes the training procedure more complicated. Eventually, we could proceed to regularize the loss function in order to avoid a near-constant prediction. However, we suspect that because the signal is constantly changing direction, the inherent delay in the forecast would not make the predictions trustworthy.

The second alternative approach suggested in Sect. 4.2 is to apply a strategy that follows the current spread trend, and uses the spread forecast as a proxy to detect the trend reversion in advance, at which point the position is exited. Two problems arise with the application of this strategy. The first is the fact that not even all spreads are smooth enough to allow the identification of local trends. The second problem is that even for those pairs for which the spread describes the desirable behaviour, the spread variation is so small that a simple delay in detecting the trend reversion is sufficient to cause a relatively large drop in the accumulated profits.

Appendix 1

This appendix includes an illustration of the resulting application of t-SNE to the clusters formed in the periods of January 2012 to December 2014 (Fig. 6.7) and January 2013 to December 2015 (Fig. 6.8). The period of January 2014 to December 2016 was considered for analysis in Sect. 6.2.3.2 (Fig. 6.1).

Fig. 6.7 OPTICS clustering in the Jan 2012–Dec 2014 formation period

Appendix 2

This appendix contains some relevant information regarding the forecasting performance measures used in this work.

Various performance measures are proposed in the literature to estimate forecast accuracy and compare different models. In this work, we analyse the results with respect to three measures. They are a function of the actual time series, which is represented by y_t, and of the forecasted values, f_t. We represent the prediction error at point t by $e_t = y_t - f_t$.

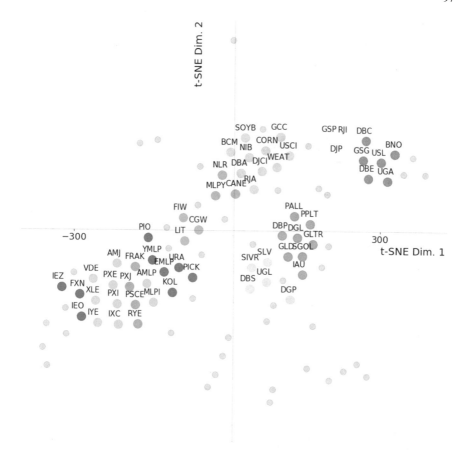

Fig. 6.8 OPTICS clustering in the Jan 2013–Dec 2015 formation period

Mean Squared Error

The MSE is defined as

$$MSE = \frac{1}{n} \sum_{t=1}^{n} e_t^2.$$ (6.1)

MSE is a measure of the average squared deviation of the predicted values from the real values. The application of the square function is particularly relevant to avoid that the positive and negative errors cancel each other out. This comes at the expense of penalizing larger errors more than small errors. Moreover, MSE is not as intuitive and easily interpretable as other measures, such as MAE.

Mean Absolute Error

The mean absolute error, MAE, is defined as

$$MSE = \frac{1}{n} \sum_{t=1}^{n} e_t^2. \tag{6.2}$$

MAE measures the average absolute deviation between the prediction and target values. As in MSE, the positive and negative deviations do not cancel out, in this case because the absolute value is taken. However, this comes at the expense of having an undefined derivative at 0. This makes the MAE less amenable to the techniques of mathematical optimization. While to optimize the MSE, one can just set the derivative equal to 0 and solve, optimizing the absolute error often requires more complex techniques.

Root Mean Squared Error

The Root Mean Squared Error corresponds to the square root of the MSE, calculated as

$$RMSE = \sqrt{MSE} = \sqrt{\frac{1}{n} \sum_{t=1}^{n} e_t^2}. \tag{6.3}$$

With the introduction of the square root, the error transforms into the same scale as of the targets, making its interpretation easier. It is worth noticing that both MSE and RMSE are similar in terms of their minimizers, meaning that every minimizer of MSE is also a minimizer of RMSE and vice versa, since the square root is a non-decreasing function. As MSE is mathematically simpler, it is more commonly used during the optimization process.

Appendix 3

This appendix contains the results of the configurations implemented for each forecasting algorithm. It is important to point out that the forecasting errors presented correspond to the average error across the five pairs being studied in this work. The model configurations which achieved the best validation mean squared error are emphasized in bold. The motive for considering the mean-squared error in the validation set for comparison purposes is justified in Sect. 6.3.3.

The results obtained for the different configurations of the ARMA model can be found in Table 6.15. Table 6.16 presents the forecasting scores obtained when using the LSTM forecasting model. Finally, the results concerning the LSTM Encoder Decoder are divided across Tables 6.17 and 6.18.

Table 6.15 ARMA forecasting results

Configuration ID	1	2	3	4	5	6	7	8	9
Parameters									
AR order (p)	1	2	2	4	5	5	**8**	10	12
MA order (p)	1	1	4	2	2	4	**3**	5	4
Validation score									
MSE(E-03)	1.542	1.513	1.512	1.509	1.511	1.511	**1.508**	1.509	1.509
RMSE (E-02)	3.032	3.007	3.006	3.004	3.006	3.006	**3.004**	3.004	3.004
MAE (E-02)	1.687	1.776	1.781	1.780	1.781	1.782	**1.780**	1.780	1.780
Test score									
MSE (E-03)	2.384	2.286	2.275	2.270	2.271	2.271	**2.264**	2.265	2.264
RMSE (E-02)	3.414	3.357	3.347	3.343	3.343	3.343	**3.339**	3.339	3.338
MAE (E-02)	1.877	1.965	1.968	1.966	1.967	1.967	**1.964**	1.964	1.964

Table 6.16 LSTM forecasting results

Configuration ID	1	2	3	4	5	6	7	8
Parameters								
# Inputs	12	12	12	12	24	**24**	24	24
# Hidden Layers	1	1	1	2	1	**1**	1	2
# Hidden Nodes	10	20	30	10-10	40	**50**	60	10-10
Validation score								
MSE (E-03)	2.73	8.53	2.18	3.03	1.80	**1.69**	1.91	2.65
RMSE (E-02)	3.73	6.29	3.63	4.52	3.30	**3.28**	3.43	4.14
MAE (E-02)	2.65	2.90	2.50	3.07	1.97	**2.04**	2.03	2.71
Test score								
MSE (E-03)	4.99	12.1	8.64	21.2	8.43	**3.35**	3.54	23.5
RMSE (E-02)	4.74	7.04	6.45	10.5	5.96	**4.30**	4.61	10.8
MAE (E-02)	3.63	5.73	5.21	82.0	4.47	**3.08**	3.36	8.06

Table 6.17 LSTM Encoder Decoder forecasting results (1)

Configuration ID	1	2	3	4	5	6	7
Parameters							
# Inputs	12	12	12	12	12	12	12
# Encoder Nodes	5	10	10	15	20	30	50
# Decoder Nodes	10	5	10	15	20	30	50
Validation score (t+1)							
MSE (E-03)	2.92	2.55	2.31	2.27	2.32	2.03	2.01
RMSE (E-02)	4.31	4.03	3.83	3.82	3.83	3.60	3.57
MAE (E-02)	2.54	2.43	2.26	2.21	2.20	2.18	2.13
Validation score (t+2)							
MSE (E-03)	2.93	2.53	2.58	2.39	2.96	2.43	2.28
RMSE (E-02)	4.41	4.06	4.03	3.94	4.28	3.94	3.88
MAE (E-02)	2.77	2.65	2.47	2.49	2.51	2.51	2.55
Test score (t+1)							
MSE (E-03)	24.2	13.2	18.2	14.4	7.46	5.72	5.48
RMSE (E-02)	10.7	8.31	9.22	8.37	6.49	5.75	5.61
MAE (E-02)	8.05	6.34	6.89	6.29	4.84	4.31	4.21
Test score (t+2)							
MSE (E-03)	19.8	11.8	11.4	23.2	10.2	8.45	8.97
RMSE (E-02)	9.85	8.18	7.92	10.1	7.43	6.91	7.17
MAE (E-02)	7.77	6.66	6.30	7.81	5.87	5.52	5.76

Table 6.18 LSTM Encoder Decoder forecasting results (2)

Configuration ID	8	9	10	**11**	12	13	14
Parameters							
# Inputs	24	24	24	**24**	24	24	24
# Encoder Nodes	5	10	10	**15**	20	30	50
# Decoder Nodes	10	5	10	**15**	20	30	50
Validation score (t+1)							
MSE (E-03)	6.45	2.73	2.08	**1.71**	2.11	1.96	2.29
RMSE (E-02)	5.83	4.04	3.54	**3.32**	3.63	3.56	3.78
MAE (E-02)	2.55	2.31	2.33	**2.13**	2.11	2.21	2.10
Validation score (t+2)							
MSE (E-03)	2.38	3.65	2.21	**2.05**	2.49	2.42	2.49
RMSE (E-02)	4.00	4.66	3.63	**3.60**	3.95	3.92	3.98
MAE (E-02)	2.49	2.68	2.45	**2.45**	2.42	2.51	2.39
Test score (t+1)							
MSE (E-03)	9.34	8.45	17.1	**4.31**	5.05	6.06	9.81
RMSE (E-02)	7.10	6.75	8.06	**5.21**	5.47	5.95	7.04
MAE (E-02)	5.53	4.79	5.65	**3.94**	4.07	4.50	5.32
Test score (t+2)							
MSE (E-03)	10.7	9.52	13.1	**9.03**	7.29	8.73	16.7
RMSE (E-02)	7.66	7.26	7.97	**7.50**	6.55	7.04	8.84
MAE (E-02)	6.22	5.36	5.38	**5.91**	5.09	5.57	6.88

References

1. Dunis CL, Laws J, Evans B (2006) Modelling and trading the gasoline crack spread: a non-linear story. Deriv Use Trading Regul 12(1–2):126–145
2. Dunis CL, Laws J, Middleton PW, Karathanasopoulos A (2015) Trading and hedging the corn/ethanol crush spread using time-varying leverage and nonlinear models. Eur J Financ 21(4):352–375
3. ETFcom (2019) ETFs - Tools, Ratings, News. https://www.etf.com/etfanalytics/etf-finder, Accessed 2019-06-30
4. ETFcom (2019) GCC Overview. https://www.etf.com/GCC#overview, Accessed 2019-07-28
5. Gers FA, Eck D, Schmidhuber J (2002) Applying lstm to time series predictable through time-window approaches. In: Neural Nets WIRN Vietri-01. Springer, Berlin, pp 193–200
6. Maaten Lvd, Hinton G (2008) Visualizing data using t-SNE. J Mach Learn Res 9(Nov):2579–2605
7. Olivier A, Jean-Luc Z, Maurice M (1993) Identification and prediction of non-linear models with recurrent neural network. In: Mira J, Cabestany J, Prieto A (eds) New trends in neural computation. Springer, Berlin, pp 531–535
8. Siami-Namini S, Tavakoli N, Namin AS (2018) A comparison of ARIMA and LSTM in forecasting time series. In: 2018 17th IEEE international conference on machine learning and applications (ICMLA), IEEE, pp 1394–1401
9. Tenti P (1996) Forecasting foreign exchange rates using recurrent neural networks. Appl Artif Intell 10(6):567–582

Chapter 7
Conclusions and Future Work

7.1 Conclusions

We start by revisiting the first research question: "Can Unsupervised Learning find more promising pairs?". If an investor is focused on obtaining the highest possible return regardless of the incurred risk, searching for all pair combinations turns out particularly appealing. But when risk is taken into consideration, the OPTICS based strategy proves more auspicious. It is capable of generating the highest average portfolio Sharpe ratio of 3.79, in comparison with 3.58 obtained when performing no clustering or 2.59 when grouping by category. Also, it shows more consistency with respect to the portion of profitable pairs in the portfolio, with an average of 86% profitable pairs, against 80% when grouping by category and 79% when performing no clustering at all. At last, it achieves more steady portfolio drawdowns, maintaining the MDD values within the acceptable range even when the other two techniques display considerable deviations. Therefore, to respond to the initial question, we may conclude that Unsupervised Learning is able to identify more promising pairs.

We proceed to revisit the second research question this work aims to answer: "Can a forecasting-based trading model achieve a more robust performance?". The results indicate that if robustness is evaluated by the number of days the portfolio value does not decline, then the proposed trading model does provide an improvement. During the period of January 2018 to December 2018, the forecasting-based models have a total of 2 (LSTM), 11 (ARMA) and 22 (LSTM Encoder-Decoder) days of portfolio decline, in comparison with the considerably more numerous 87 days, obtained when using the standard threshold-based model. However, the overall portfolio profitability is reduced as a side effect. Even so, we find evidence that the 10-year period required to apply the forecasting-based model significantly harms the performance, suggesting its profitability could improve if used in a more convenient period. Thus, to answer the second research question, we may assert that although the forecasting-based model showed potential if applied in a more convenient setting, the decline in profits observed in these conditions does not compensate the increased robustness.

© The Author(s), under exclusive license to Springer Nature Switzerland AG 2021
S. Moraes Sarmento and N. Horta, *A Machine Learning based Pairs Trading
Investment Strategy*, SpringerBriefs in Computational Intelligence,
https://doi.org/10.1007/978-3-030-47251-1_7

Some additional contributions of this work are:

- Demonstrating that the application of Pairs Trading using commodity ETFs in an intra-day setting can be beneficial.
- Proving Pairs Trading can be still profitable after transaction costs in recent years.
- Demonstrating the Hurst exponent can be used to enforce mean-reversion.
- Evaluating three different heuristics for building portfolios of pairs and finding that selecting pairs based on the validation performance is advantageous.

7.2   Future Work

Several possible directions of study could be taken as a follow-up of this work. It may be interesting to continue exploring the proposed forecasting-based trading model (1) or to define different schemes for a Pairs Trading strategy (2).

(1) Improving the proposed forecasting-based model:

- Add more features to predict the price variations rather than constraining the features to lagged prices.
- Increase the data frequency (e.g. 1-min frequency) to enable a reduction in the required formation period and consequently find more pairs. It should be noted that this will be very demanding from a computational point of view.
- Experiment training the Artificial Neural Networks in a classification setting. Although regression is seemingly the most obvious solution, given that the output being predicted can take any real value, classification might bring other advantages. The reason is that the MSE loss is much harder to optimize than a more stable loss such as Softmax, and it is less robust since outliers can introduce huge gradients. Classification has the additional benefit of providing a distribution over the regression outputs, not just a single output with no indication of its confidence.

(2) Other approaches:

- Construct a metric to rank pairs in the portfolio, instead of applying an equal weighting scheme. For example, based on a similarity measure between the securities' features resulting from the application of PCA, or something more empirical, such as the relative performance in the validation period.
- Explore the market conditions in which the proposed strategy generates more returns, and investigate whether the investor can benefit from predicting such scenarios in advance.
- Combine commodity-linked ETFs and other security types (e.g. futures) in the same pair and see if the investor can benefit from the additional expected volatility.

Printed in the United States
By Bookmasters